Synthese Library

Studies in Epistemology, Logic, Methodology, and Philosophy of Science

Volume 526

The aim of *Synthese Library* is to provide a forum for the best current work in the methodology and philosophy of science and in epistemology, all broadly understood. A wide variety of different approaches have traditionally been represented in the Library, and every effort is made to maintain this variety, not for its own sake, but because we believe that there are many fruitful and illuminating approaches to the philosophy of science and related disciplines.

Special attention is paid to methodological studies which illustrate the interplay of empirical and philosophical viewpoints and to contributions to the formal (logical, set-theoretical, mathematical, information-theoretical, decision-theoretical, etc.) methodology of empirical sciences. Likewise, the applications of logical methods to epistemology as well as philosophically and methodologically relevant studies in logic are strongly encouraged. The emphasis on logic will be tempered by interest in the psychological, historical, and sociological aspects of science. In addition to monographs *Synthese Library* publishes thematically unified anthologies and edited volumes with a well-defined topical focus inside the aim and scope of the book series. The contributions in the volumes are expected to be focused and structurally organized in accordance with the central theme(s), and should be tied together by an extensive editorial introduction or set of introductions if the volume is divided into parts. An extensive bibliography and index are mandatory.

Guilherme Ghisoni da Silva

The Philosophical Foundations of Photography as a Means of Communication

Springer

Guilherme Ghisoni da Silva
Faculty of Philosophy
Federal University of Goiás
Goiânia, Brazil

ISSN 0166-6991 ISSN 2542-8292 (electronic)
Synthese Library
ISBN 978-3-032-02648-4 ISBN 978-3-032-02649-1 (eBook)
https://doi.org/10.1007/978-3-032-02649-1

This Springer imprint is published by the registered company Springer Nature Switzerland AG.
The registered company address is: Gewerbestrasse 11, 6330 Cham, Switzerland

To my beloved parents, Zélia and Rolando.

Preface

This book arises from the confluence of two themes that have long been dear to me: photography and the work of philosophers dedicated to understanding language and its uses. My involvement with photography, as an aspiring professional photographer, preceded my academic study of philosophy. In fact, it was the study of authors exploring themes in the philosophy of photography that led me to an academic career in philosophy. But ever since I began exploring the philosophy of photography and adjacent topics, I have always felt the absence of a research field focusing on the contextualized role of photographs as part of a communicative intention. Works in the philosophy of photography, particularly by authors in the analytic tradition, have often been excessively focused on issues concerning the causal origins of photography. Related discussions have primarily dealt with pictorial content, seeking to explain how two-dimensional surfaces can enable us to experience a scene composed of objects arranged within a three-dimensional pictorial space. There was also an excessive restriction of the topic of photography to discussions about its status as an art form, addressing the longstanding problem of whether photography can be art. Little to no attention was given to the everyday uses of photography and how we use it to *communicate*, a use that is increasingly common in our digital daily lives.

The certainty that a field of study was missing in the philosophy of photography came to me through two experiences, whose personal significance had diametrically opposite values. For more than ten years, I have participated as a visual artist in the gallery circuit, presenting my photographic works in national and international exhibitions. I have also collaborated with renowned Brazilian photographers, working as an art critic and contributing to the writing and editing of chapters and books. This dual citizenship, as both an artist and a philosopher, led me to create a Research Laboratory for the Philosophy of Photography at the Federal University of Goiás (in Brazil), which was supported by the Goiás Research Support Foundation (FAPEG) and the National Council for Scientific and Technological Development (CNPq). Since the formal establishment of the Laboratory in 2016, I have collaborated with various artists, photographers, and students, contributing to the development of both visual and academic research through the study of philosophical themes, particularly the philosophy of photography. This experience, besides being extremely gratifying and personally fulfilling, has allowed me to gain a practical understanding of artistic creation processes and the relationship between photographs and their context of use. Photography is not merely the photographic picture, detached from its use; it communicates based on the social context in which it is used, in relation to the biography of the picture's user. What was lacking was a theory that explained how this communication was structured in relation to this broader concept of context. To address this, it was necessary to deepen the distinction between what the photograph *depicts* and what its use *communicates* and to understand how we transition from photograph to communication.

The other experience that confirmed to me the absence of a field of study in the philosophy of photography was the research I conducted between 2019 and 2020, during which I monitored for nine months the visual production of more than 300 far-right WhatsApp groups from Brazil. Through a digital platform created by the Federal University of Minas Gerais (Brazil), I had daily access to the thousands of pictures and memes distributed in these groups (and I am grateful to my colleague Fabrício Benevenuto for granting me access to the *WhatsApp Monitor*). It was a psychologically challenging task, as I was constantly exposed to different forms of violence perpetrated through the use of these pictures. The philosophical problem posed by these photographs was that many did not contain violent pictorial content, yet they were used with the intent to communicate ideas and positions that promoted extremism, misogyny, racism, xenophobia, and a systematic assault on minorities and individuals. Understanding these forms of violence required a clearer distinction between what the picture depicts and what its use communicates. Given that what was communicated often violated constitutional principles and individual rights, it was crucial to establish a theoretical foundation for communication via photographs to clearly describe the crimes committed and to ensure accountability for those forms of violence. However, to my surprise, such a theoretical foundation was scarce, or it led to theoretical labyrinths that precluded practical application. As a result, these crimes remain unpunished, as in many cases we lacked the ability to clearly formulate what the crime was. The path I found to partially address this gap, in the article published under the title "Memes War: The Political Use of Pictures in Brazil 2019" (2020), was to explore the parallel between photographs and speech acts. After working on this paper, I became certain that a new theoretical starting point was necessary to serve as the foundation for constructing a theory of photographic communication. To achieve this, it would be necessary to first understand the most basic communicative uses of photography, setting aside approaches that prioritized photography as art.

At the time of conducting the abovementioned research, I was a visiting professor at the University of British Columbia, in Vancouver, Canada. There, I had the opportunity to thoroughly explore the work of Dominic Lopes, with whom I was able to interact. Lopes' vast body of work is extremely fruitful and brings numerous contributions to picture theory, theory of pictorial expression, philosophy of photography, philosophy of art, aesthetics, among other fields. To my delight, I found in his recent thought a critique of the Orthodox philosophy of photography and a proposal for a "social turn," which could serve as the turning point for the theory I sought to construct. The fundamental idea of Lopes' "social turn" is that photography is divided into different social practices, each with distinct purposes and rules that determine the epistemic status of the photographic picture within the practice. Some practices are epistemic, while others are non-epistemic. Through the study of John Searle's social ontology, I found elements to connect Lopes' "social turn" to the idea that we use photographs to perform speech acts. The study of the seminal works of David Novitz and Rae Langton, among other authors, allowed me to deepen the idea that we use pictures to perform speech acts and to develop the notion of a *photographic illocutionary act*, in which what the picture depicts plays the role of the *locutionary act*. The next step was the development of a taxonomy of photographic speech acts, which would also serve as the basis for a taxonomy of social practices, and the understanding of how different practices have distinct conditions as constitutive rules. Drawing on Korta and Perry, I was able to distinguish two fields of study that take into account the role of context in determining communication through photographs: *near-side pictorial pragmatics*, concerned with the nature of certain facts that are relevant to determining what is depicted, and *far-side pictorial pragmatics*, focused on what happens beyond depiction (what speech acts are performed in or by using the image, or what implicatures are generated by photographically depicting what is depicted).

What I termed *near-side pictorial pragmatics* had already been sufficiently mapped by image theories and excellent recent contributions. My main task would be the development of a *far-side pictorial pragmatics* applied to photographic communication. The creation of this second field was accomplished by developing an ontology of photographic social practices in conjunction with speech act theory, as a means of advancing Lopes's "social turn."

The reader will find in this book the journey from Orthodox philosophy of photography to the development of an ontology of social photographic practices, with particular attention to the way in which different photographic speech acts require the fulfillment of distinct conditions. It is through the contextualized use of photographic pictures within distinct social practices that photography is often employed as a means of communication. Thus, this journey constitutes the construction of the philosophical foundations for a communicational theory of photography, oriented toward future developments. I hope, with this, to have taken a decisive step toward building and consolidating a theory of ordinary photographic practices, allowing for an understanding of how we communicate through the use of photography. I also hope that future refinements and further explorations will be put forward by other researchers. The ideas presented here are merely a new starting point.

I am grateful to all the participants of the Research Laboratory for the Philosophy of Photography, with whom I had the opportunity to discuss and repeatedly refine the ideas presented in this book. I am also grateful to all the institutions that supported the development of the Laboratory. I am deeply indebted to the renowned Brazilian photographer Claudio Edinger, who embraced my visual work and introduced me to the world of galleries and exhibitions, and with whom I had the opportunity to participate in several group exhibitions and collaborate on various books. It was the access to the *artworld* as an artist, provided by Edinger's support of my visual work, that gave me a broader understanding of photographic practices and how photography serves as a form of communication. I am grateful beyond words to Dominic Lopes, who made possible my time in Vancouver, during which I developed the fundamental ideas of this book. Lopes' brilliance and good-heartedness were crucial in enabling me to move beyond my exegesis-based training in the history of philosophy (a common restriction still present in Brazil) and find my own voice. I am grateful to all my colleagues who directly or indirectly contributed to the development of the ideas presented here. I am grateful to the National Council for Scientific and Technological Development (CNPq), from Brazil, for supporting my research with a research grant. My deepest gratitude goes to the University of Goiás, which has accommodated me since 2011, and to all the professors, students, and staff of the institution.

I am grateful to my family and to my wife, Celira Botelho Mendes, without whom this work would not have been possible.

Goiânia, Brazil
September 5, 2024

Guilherme Ghisoni da Silva

Introduction

The main concern of this book is to lay the conceptual foundations of a theory of photography as a means of communication. Photography has increasingly played a central role in our lives. Every day, we are exposed to thousands of photographs, and we are progressively relying on them as a means of communication. All of us have smartphones at hand, allowing us to send and receive pictures that travel across the globe almost at the speed of light. Much of what we know about the world and what is communicated to us is conveyed through photographs. Moreover, what we communicate to those with whom we interact is increasingly established by the photographs we send through apps or post on social media. There is always a communicative intention behind these actions. The use of photographic pictures has become the *lingua franca* of our visual communication.

Unfortunately, the significance of photography as a means of communication is inversely proportional to the reduced attention it has received in academic studies, particularly within the analytic tradition of the philosophy of photography. This neglect is largely due to the way Orthodox philosophy of photography has focused on themes that contribute little or nothing to understanding its everyday uses, prioritizing debates over a supposed essence of photography, determined by its causal genesis, and the challenges this poses to understanding photography as art. Moreover, the central problem, one that would block the development of a theory of photographic communication even before its birth (rendering this book's project stillborn), is that the theory of communication adopted by several Orthodox authors undermines the idea that the use of photographs constitutes genuine communication. A critical analysis of the Orthodox position and the overcoming of this restriction will be necessary starting points for the development of the theory this book aims to construct.

I will take Robert Hopkins as the paradigm of Orthodox theories, as I consider it the most fruitful and interesting version. Hopkins, like Kendall Walton and Roger Scruton, accepts the general lines of Paul Grice's theory of communication. Grice's theory, while influential, limits genuine communication to the use of non-natural meaning, thereby excluding photographs as a form of communication since photography is regarded as a model example of *natural meaning*. It is due to the adoption of Grice's theory that Hopkins argues in favor of the communicative poverty of photography as an art form. Thus, in order to render the project of this book viable, a critical analysis of Grice's theory will be necessary, along with the rejection (as proposed by François Recanati, Daniel Sperber, and Deirdre Wilson, among others) of the thesis that imposes drastic restrictions on the idea of photographic communication. As a result of this critique, we will reach a broader perspective on photographic communication, in line with Sperber and Wilson's theory, in which communication can be thought of as composed of two layers: *a foundational layer of information* about any subject and a *second layer indicating that the first layer was intentionally manifested.*

Another point of departure from Orthodoxy will be through the adoption of the New Theory of the philosophy of photography, developed by Paloma Atencia-Linares, Dominic Lopes, Catharine Abell, and Diarmuid Costello (drawing on Patrick Maynard and Dan Wilson (née Phillips)). In Lopes' version, we find the idea that photography is constituted by different *social practices* (which I will use as a central thesis for the development of this book). The way Orthodox thinkers approach photography would be a prioritization of a certain type of use at the expense of other possible practices. The "social turn" also provides us with a path beyond the limitations of the debate to the context of art, as there are several non-artistic photographic social practices.

I will also abandon the way Hopkins formulates photographic communication in relation to the elements available to the picture *producer* for communicating something to the *spectator*. A photograph is not always used for the same purpose for which it was produced. Much of our everyday communication happens through the intentional use of photographic images that were not produced by us. For these reasons, I will prioritize *user–spectator* communication, considering *producer–spectator* communication as a special case in which the user is the producer.

But I do not believe my position should be read as a refutation of Hopkins' theory, or of Orthodox authors in general. Philosophy, especially within the analytic tradition, has often been structured as a kind of theoretical "joust," where the aim is to "knock the opponent off their horse" by refuting their arguments and ideas. I find this approach somewhat unproductive. Hopkins' position, in my view, is correct, but only for a very limited class of cases, where photographs are used in accordance with the supposed essence of photography, in artistic contexts, and under a restrictive reading of Grice's theory of communication. Expanding the scope of our analysis will allow us to adopt some of Hopkins' claims about photographic communication, while remaining aware of their limitations. In other words, in a general sense, I will adopt a methodology that, rather than seeking to "knock the opponent down," focuses on understanding which aspects of their theories can be integrated into the development of my own, even when there are key differences between our positions.

Based on the understanding of communication as structured in two layers, I will deepen the analysis of photographic communication by drawing a parallel with *speech act theory*, developed by John Austin and John Searle. The foundational idea is that the pictorial content of a figurative photographic image, context-sensitively determined, could serve as a *locutionary act* (constituting the foundational layer of information). The intentional use of the image, in turn, would constitute an *illocutionary act* (a second layer signifying intentional use).

The idea that we use pictures to perform speech acts is not new, and it finds support in the seminal works of authors such as David Novitz and Rae Langton. Regarding the parallel between pictures and speech acts, the development I propose in this book involves a deeper understanding of the specific case of photography. I will distinguish five ways in which a photograph can relate to a speech act, prioritizing some of them, with the possibility that the use of the same image may perform multiple acts simultaneously or indirectly. I will argue, based on Langton, that the role of photography is not limited to determining a propositional content to which an illocutionary force is added (either verbally or as a presupposed illocutionary force in the context of use), but the contextualized use of the photograph, depending on what it depicts, how it depicts it, and how it is used, can also determine the illocutionary force of the act. Furthermore, drawing on some of the central ideas from Searle, I will propose a taxonomy of photographic illocutionary acts, with special attention to the idea that in different illocutionary acts, photography will have different *directions of fit* in relation to the world. I will argue that photographs can perform all five basic categories of speech acts proposed by Searle, allowing photographic illocutionary acts to be distinguished as *assertive*, *directive*, *commissive*, *expressive*, and *declaratory*.

The idea that photographic communication is composed of two layers will allow me to separate two fields of study within the philosophy of photography and picture theory, both of which pay attention to the context-sensitive nature of pictorial communication: one field concerned with the nature of certain facts relevant to determining what is depicted, which I will call (adapting some ideas of Korta and Perry) *near-side pictorial pragmatics*, and a broader field focused on what happens beyond depiction, which I will call *far-side pictorial pragmatics*. The first field of study will be devoted to picture theories and pictorial reference, aiming to understand how the use of a picture as a foundational layer of information in communication requires that the user and the spectator be in sufficient agreement about pictorial competence and the conceptual requirements necessary to determine the propositional content related to the picture's content (topics we will address in Chap. 2). Attention to the expressiveness of the picture will also be important, as there are significant context-sensitive aspects in determining the mental states, feelings, and emotions a picture expresses. The second field will be dedicated to understanding what *speech acts* are performed by or through the use of the picture, or what *implicatures* are generated by photographically depicting what is depicted. The creation of a *far-side pictorial pragmatics* applied to photography constitutes the central contribution of this book.

The importance of context sensitivity, both in terms of what a picture depicts and what its use communicates, will be explored through *four types of underdeterminations* (three of which are constitutive of all pictures and one specific to photography). The same arrangement of colors and shapes on a two-dimensional surface could determine different pictorial contents and, consequently, different propositional contents in different contexts of use. I will refer to this idea as *pictorial underdetermination*. Furthermore, the same pictorial content could be mapped onto the domain of existing objects in different ways. Although photography often prioritizes the causal mechanism of reference, I will argue that other mechanisms may be at play; thus, allowing the use of photographs to determine propositional contents with different levels of quantification. I will call this second case *denotative underdetermination*. The third case concerns the fact that visually identical photographs can be produced by different means, using distinct processes and photographic technologies. Thus, two identical photographs may have different epistemic status and distinct functions and powers within the same social practice or across different social practices. I will refer to this as *photographic underdetermination*. The fourth type of underdetermination relates to the possibility that we could use the same photograph, with the same content, denotation, and means of production to perform different speech acts. I will refer to this fourth case as *communicational underdetermination*. The central point that underscores the importance of the contextualized use of images and the *user-spectator* relationship is that overcoming these underdeterminations will only occur through contextualized use, where the picture's user and viewer agree on the elements that determine what the image depicts, denotes, and how it was produced and used to communicate.

The view that communication must be approached by considering the context of use and the relationship between user and spectator will lead to an important shift in perspective, one that sidesteps some of the main issues in the philosophy of photography and picture theory, or at least compels a significant reformulation of these questions. Instead of asking whether photographs can be true or false, the communicative perspective adopted here will lead us to frame the problem as a question of whether we *use* photographic images in contexts where we judge the photograph in relation to truth or falsity. Instead of asking whether images have propositional content, the perspective adopted here will lead us to reformulate the question as whether we *use* photographic images to determine propositional contents. Rather than asking whether photographs perform speech acts, we should ask whether we *use* photographic images with a communicative intention in which the use of the photograph performs speech acts. From this perspective, which places emphasis on how we *use* photographs in communicative interlocutory contexts, such questions will certainly receive positive answers. Understanding how these answers are articulated and unfold will play a central role at various moments in the development of this book.

A key aspect of the ordinary use of photographs is the categorization of photographs in relation to different social practices of use. We say that one photograph is photojournalistic, another advertising, one picture is artistic, another forensic, one scientific, and yet another serves as a diagnostic medical image, among other possible uses. But what role does this categorization play in determining how we use these pictures to communicate? What difference does it make if we situate a photograph in relation to one social practice of use or another?

The answers to these questions will be reached through the development of an *ontology of social photographic practices*, in which I will connect the "social turn" proposed by Dominic Lopes to John Searle's social ontology. I consider the development of this ontology of social photographic practices to be the main contribution of this book (in Chap. 4). But it is important to note that this book is not about the foundations of social ontology. In fact, I do not take a definitive stance on many of the fundamental questions in the field. Nor is it my intention to claim that *all* photographic social practices exhibit the structure I attribute to practices in this book. What matters for my purposes is the conviction that it is fruitful to approach the main photographic practices that structure our everyday lives through the conceptual framework of social ontology adopted here.

The starting point of the theory proposed here is the idea that the communicative uses of photography take place within distinct social practices, in which participants share a *collective intentionality* that determines the *perceived function* of the practice. Thus, photographic social practices will have different functions that establish different *purposes* to which photographs are used within the practice. The function of a practice frequently plays a teleological role in the development of photographic technologies and brings with it the possibility of a theory of value, since we tend to place greater value on photographs that enhance the institutionalized pursuit of their functions. Historically consolidated practices, which I will classify as *conventional contexts*, typically have institutions and rules that must be followed for an image to be used within the practice and its corresponding institutions. It is through compliance with the *constitutive rules* of a practice that the assignment of a *status function* to photographs occurs; expressed in the classification of a photograph as photojournalism, advertising, artwork, forensic, scientific, etc. The assignment of a status function through the fulfillment of the constitutive rules of the practice to a photograph may also confer upon the picture *positive* and/or *negative powers* (including deontic or epistemic powers), such as rights and duties, within that practice. These are the elements that structure photographic communication within conventional contexts.

At this point in the analysis, the taxonomy of photographic speech acts will serve a dual function. Since the purpose to which photographs are directed within photographic social practices can be understood as the speech act (or acts) that the photographs primarily perform within the practice, we can understand this purpose as the photograph's *illocutionary purpose*. The taxonomy of photographic speech acts will thus provide the foundation for a basic taxonomy of photographic social practices.

An important part of the final analyses of this book (in Chap. 4) will involve understanding the rules that practices impose on the use of photographs, based on the conditions for the proper implementation of the speech act (or acts) that constitute the purpose of the practice. Special attention will be given to what can be described, following Searle, as *preparatory conditions* and *content conditions*. Preparatory conditions specify the requirements of authority and institutional affiliation necessary for participation in a practice. Content conditions, on the other hand, can be understood in photographic terms as the determination of what the picture should depict and the manner in which it should be produced. In some contexts, only images that depict objects in a certain way, that conform to specific aesthetic standards, and that are produced in a particular manner will be accepted. A contribution I consider significant in this book is the detailed analysis of different types of conditions for the proper implementation of photographic speech acts.

The fulfillment of these conditions is what determines the epistemic value of the picture in the case of assertive practices, such that its status function may allow for factive inferences of the form: "If the photograph X has the *status function* S in the practice P, then the fact that X means Y implies Y." We can thus conceive of the factive experience afforded by some photographs as related to an epistemic power that results from the attribution of a status function within assertive practices, following the very model of *deontic powers* that Searle (2010) associates with the assignment of status functions. In other words, the epistemic power to afford a factive experience does not stem from photography itself, but from the *rule-governed* use of pictures within certain social practices, which enables an institutional recognition that confers upon the picture a normative role.

This approach will allow us to draw an important distinction between contexts that can be described as *conventional* and those that are *non-conventional*. Although communication will be conceived here as centered on the *manifestly intentional use* of a photograph, in conventional contexts, we can adopt a non-intentionalist stance on communication, since intending the illocutionary purpose of the social practice will be a prerequisite for participation (an *essential condition*, in Searle's terminology). Thus, the mere presence of a photograph within a practice already permits the attribution of an intentionality aligned with the practice's purpose, without the need for the user to manifest this intention overtly. The presence of the picture in a conventional context also carries the presupposition that the other prerequisites for participation have been met.

However, we also use photographs for communication in contexts where there is no pre-determined illocutionary purpose for which the image is intended, nor the presumption that certain requirements have been met. I will refer to these cases as *non-conventional contexts*. Communication in non-conventional contexts will be centered on a *reflexive intention*: the overt manifestation of the user's communicative intention and the viewer's recognition of that intention. In this sense, the return to Grice's theory, modified to accommodate photographic communication, will play an important role. It will also be important to understand how photographic communication is based on the Cooperative Principle and a photographic version of communicative Maxims.

But in our everyday life, what we encounter are practices with varying degrees of regulation, forming a *continuum* between the theoretical paradigms of conventional and non-conventional contexts. And since in conventional contexts there is still the possibility that the use of a picture may perform secondary or indirect speech acts, there remains room for communication to be partially centered on the ostensive manifestation of a communicative intention, even in conventional contexts.

It is also important to note that the distinction between what a picture depicts and what it communicates will allow for a bipartite perspective on truth. A photograph that visually depicts its causal referent accurately, and could thus be accepted as true and genuine from a pictorial standpoint, can be used to perform an assertive speech act that may be false, and *vice versa*. Understanding how truth relates to photographic images in different ways and with varying levels of quantification will also be an important task of this book.

The book will be structured as follows. In Chap. 1, I will focus on analyzing the Orthodox theory, particularly Hopkins' version (Sect. 1.1), and the critiques that will allow us to broaden our understanding of photographic communication (Sect. 1.2). To do so, we will explore Paul Grice's theory of communication assumed by Hopkins (Sect. 1.3) and deepen our understanding of Orthodox Theories (Sect. 1.4). At the end of the chapter, we will turn to the study of the New Theory (Sect. 1.5) and use the preceding analyses to propose the foundations of a broader theory of photographic communication (Sect. 1.6).

I will begin Chap. 2 by drawing a distinction between *near-side* and *far-side pictorial pragmatics*, with the aim of focusing on the former in the chapter. My analysis will be restricted to figurative pictures. The idea that context-sensitive elements determine what an image depicts and denotes will require an understanding of some fundamental aspects of picture theories and theories of pictorial reference. In Sect. 2.1, we will explore how an image's design is mapped onto the domain of meanings, determining its pictorial content, and in Sect. 2.2, we will examine the different ways in which pictorial content can be mapped onto the domain of existing objects. In Sect. 2.3, I will apply the previous discussions on pictorial reference to the case of photography, aiming to dispel some recurring misunderstandings regarding denotation in photographic contexts. Section 2.4 will be devoted to Dominic Lopes' (2006) theory of pictorial expression, since pictorial expression will play an important role in understanding expressive pictorial speech acts (in Sect. 3.5.4). At the end of the chapter, I will address the relationship between pictorial content and propositional content, with particular attention to how John Kulvicki draws parallels between David Kaplan's theory of indexicals and the context-sensitive use of images (Sect. 2.5).

Chapter 3 will be dedicated to the foundations of *far-side pictorial pragmatics*. We will begin with an analysis of John Austin's doctrine of infelicities (Sect. 3.1) and how David Novitz (Sect. 3.2.1) and Rae Langton (Sect. 3.2.2) apply Austin's ideas to pictures. Section 3.3 will focus on John Searle's taxonomy of speech acts. In Sect. 3.4, we will explore the different ways photographs can be related to speech acts. In Sect. 3.5, I will propose a taxonomy of photographic speech acts, arguing that photographs can be used to perform the five basic types of speech acts proposed by Searle. The chapter will conclude with an analysis of Paul Grice's communicative maxims, developing photographic versions of these Maxims, which will allow us to explore the notion of *photographic implicature* as a way to advance the idea of indirect speech acts (Sect. 3.6).

Chapter 4 contains the core of the book. The chapter begins with the development of an *ontology of photographic social practices*, in which I will use the taxonomy of photographic speech acts as a taxonomy of photographic social practices, since we can understand the purpose to which photographs are directed within practices as their *illocutionary purpose*. In Sect. 4.2, we will focus on the idea that different practices will have different rules for participation. I will explore the idea that the felicity conditions for fulfilling the illocutionary purpose of the practice often serve as constitutive rules. Special attention will be given to the idea of *content conditions*, which will be explored in detail in Sect. 4.3, where I will propose a distinction between pictorial rules, production rules, and aesthetic rules. The chapter will conclude with an exploration of how communication is structured in non-conventional contexts (Sect. 4.4).

It is paramount to note that this book is not an empirical study of existing social practices and how they fit into the basic taxonomy here proposed. The task of this book is philosophical, in the sense that I aim to provide the conceptual foundations for thinking about the structure of photographic social practices and photographic communication. The study of which practices exist, how the purposes of photography within those practices relate to the taxonomy, and what are the actual conditions for participation would require an empirical study that goes beyond the scope of this book. My goal is to provide a theoretical framework. Thus, what the reader will find in this book is the proposal of a starting point for the development of a new field of study on philosophy of photography, focused on how we use photographs in our daily communication. I hope that other researchers will recognize the importance of this field and how fruitful it can be, and that this work will lead to the refinement of the theses and concepts here presented.

Contents

Chapter 1
The Orthodox Theory of Photographic Communication

Abstract This chapter aims to analyze the Orthodox approach to photographic communication, using Robert Hopkins' theory as a reference. The primary goal is to identify the constraints he places on photography that led him to conclude that it cannot fully develop as a communicative representational art. The central point of the analysis will be the study of Paul Grice's theory of communication, which Hopkins adopts. There is evidence that Grice would oppose the idea that the use of photographs constitutes genuine communication. The critique of Grice's thesis, which leads to this position, will be crucial for the viability of a theory of photographic communication. It will also be important to analyze Hopkins' assumptions about the nature of photography, in order to achieve a broader perspective on photographic communication. The main path for this expansion will be the study of the New Theory of the philosophy of photography. In Lopes' version of the New Theory, we will find the "social turn," which will be one of the foundational ideas of this book. As a result of the analyses in this chapter, we will reach a broader perspective on photographic communication, in line with Sperber and Wilson's theory, in which communication can be thought of as composed of two layers: *a foundational layer of information* about any subject and a *second layer indicating that the first layer was intentionally manifested.*

Keywords Communication · Social practices · Natural meaning · Non-natural meaning · Social turn

Every general classification runs the risk of imposing distortions by overlooking nuanced cases that fall between general concepts. Nevertheless, the distinction between two major groups in the philosophy of photography, although it may overlook transitional cases, has proven both useful and recurrent. I am referring here to the approach established in the analytic tradition that distinguishes the history of the philosophy of photography between *Orthodox Theory* and the *New Theory*. The significance of this distinction lies in its ability to identify the fundamental theses

G. Ghisoni da Silva, *The Philosophical Foundations of Photography as a Means of Communication*, Synthese Library 526,
https://doi.org/10.1007/978-3-032-02649-1_1

endorsed by different groups of authors, which, in turn, shape the commitments and consequences that define these two interpretations.[1]

The purpose of Chap. 1 is to explore some of the core ideas accepted by Orthodox theories and how they shape a particular understanding of photographic communication (imposing drastic restrictions on it or even discarding it completely). To ensure that our analyses are not based on overly broad descriptions that fail to capture the specificities of individual authors, I will take Robert Hopkins' theory (2015) as a reference, as I believe it to be the most sophisticated and fruitful version of an Orthodox Theory, with important contributions for the concept of photographic communication.

Hopkins' starting point is the restriction of his analysis to what he calls *authentic photography*. This is a common move in Orthodox theories (similar to Roger Scruton's (2008) restriction of his analyses to *ideal photographs* and Kendall Walton's (2008a) focus on *transparent pictures*). Hopkins argues that authentic photography "is not able to develop to the full as a communicative representational art" (2015: 329). Our initial aim will be to understand the theses about the nature of photography and communication that lead Hopkins to his conclusion about the limitations of photography as a means of communicating artistically relevant thoughts.

Among these ideas, the acceptance of Paul Grice's theory of communication (1957) plays a crucial role, as it serves as common ground among many Orthodox theories (also present in the works of Scruton (2008) and Walton (2008a)). Grice's theory, although highly influential, confines genuine communication to the use of *non-natural meaning*, thereby excluding photographs as a form of communication since photography is regarded as a model example of *natural meaning*. Communication through the use of photographs would not be a genuine form of communication. Critically examining Grice's theory will be essential for establishing the idea of photographic communication and moving beyond the restrictions imposed by Hopkins.

Another key point of departure will be the rejection of the emphasis on a supposed *essence* of photography (which Hopkins refers to as *authentic photography*). Overcoming this restriction will be achieved through the study of the New Theory of the philosophy of photography. The New Theory offers a new understanding, centered on the idea of photographic *technology* and the acceptance that photographic technologies can be used in various ways, which may or may not align with the supposed essence of photography as described by Orthodox theories.

To avoid the problem of an overly general approach, I will focus on Dominic Lopes' version of the New Theory (2016, 2018). While several criticisms can be directed at Lopes, one of his central ideas will play a fundamental role in this book.

[1] Throughout the book, it will become clear that, despite being a work on *photography*, no pictures are included, except for Jeff Wall's photograph in Sect. 4.4. There are two main reasons for this. First, the costs involved in obtaining the rights to use certain images turned out to be prohibitive. Second, in today's digital age, it is quite easy for readers to search for images online. Therefore, when a specific photograph is mentioned, I encourage readers to look it up online to enhance their reading experience.

Lopes argues that photography is constituted by different *social practices*. Some practices are epistemic, while others are non-epistemic; some are artistic, while others are non-artistic. Different practices may impose different rules on the production of photographs. These considerations constitute the foundation of a "social turn" in the philosophy of photography, which will lead to substantial interpretative advantages over Orthodox Theory.

It will also be crucial for the purpose of this book to move beyond Hopkins' restriction of photographic communication to the artistic context and the producer-spectator perspective.

The expansion of the scope by rejecting the restrictions imposed by Hopkins will ultimately allow us to reach a broader perspective on photographic communication, in line with Sperber and Wilson's (1995) theory of communication, in which communication can be understood as composed of two layers: *a foundational layer of information* about any subject and a *second layer indicating that the first layer was intentionally manifested*. This idea will be further developed in later chapters through a parallel with speech act theory, in which what the image depicts plays the role of the locutionary act (constituting the foundational layer of information), while the intentional use of the image would, in turn, constitute an illocutionary act (a second layer signifying intentional use). This will be the path through which I will develop Lopes' "social turn" and establish the foundations of an ontology of photographic social practices.

1.1 The Limits of Photography as Communicative Representational Art in Robert Hopkins

Robert Hopkins, in a 2015 article, elects painting as the paradigm of handmade images and argues that "authentic photography, unlike painting, is not able to develop to the full as a communicative representational art" (329). Although he introduces important distinctions, his thesis echoes Roger Scruton's Orthodox position in his controversial article "Photography and Representation" (2008). Scruton posits the controversial thesis that *ideal* photography is not *representation*. Ideal photography is conceived as a logical ideal in which the relationship between the photograph and its subject is purely causal rather than intentional, rendering the image as "a record of how an actual object looked" (2008: 140). Since representation, in which x stands for y, depends on intentionality, and a photograph is not produced intentionally, ideal photography, according to the author, is not representation. And any effort to introduce intentionality into the picture would "pollute the ideal of their craft with the aims and methods of painting" (Scruton 2008: 139). Therefore, ideal photography would not be representational art, as it is not representation. This also precludes photography as a means of communication, as it would not be a "vehicle of representational thought" (2008: 161). Conversely, photography could be representation, artistic representation, and communicate thoughts, but at the cost of becoming "painting"—something that would no longer be, in essence, an *ideal photograph.*

Hopkins' formulation is precise enough to clearly delineate how his position diverges from Scruton's:

(i) His object of analysis is *authentic* photography, not *ideal* photography.
(ii) He accepts that authentic photography is *representation.*
(iii) He accepts that authentic photography can be *representational art.*
(iv) He accepts that authentic photography can be a *vehicle for communicating representational thoughts.*

His central claim is that, when comparing authentic photography to painting, photography reveals itself as limited as a vehicle for communicating artistically interesting representational thoughts. Thus, his thesis does not argue for the impossibility of photographic art, but for the limitations of photography as an artistic medium. The points that align his position with Scruton are the shift from ordinary uses of photography to the analysis of what would constitute the supposed *essence* and the resulting skepticism about photography's artistic aspirations—albeit moderate and not as radical as Scruton's. However, for the purposes of this book, Hopkins' position offers a significant interpretative advantage over Scruton's orthodox view, as it paves the way for and explores the understanding of photography as a *means of communication*, even if this medium proves to be limited compared to painting.

According to Hopkins, in authentic photography, a scene composed of objects, events, or arrays serves as input in the photographic process and causes the pictorial content via a chain of *mind-independent sufficient causes.* When the image is produced in this manner, the "imprinting" of the image occurs, which, according to him, "is central to photography's self-conception" (2015: 330).[2]

[2] This characterization of "imprinting a picture" presents an important difference compared to Kendall Walton's thesis of photographic transparency (2008a, b). Walton's stance on the distinction between photographs and paintings can be framed, as Dominic Lopes (2004, 2016) does, in the following terms: paintings are a form of *belief-dependent feature tracking*, as opposed to photographs, which are a form of *belief-independent feature tracking*. It is this independence that makes photography "transparent" in the technical sense proposed by Walton. Lopes critiques Walton by arguing that handmade images could also be belief-independent. One of his examples is the case of Nadia, a girl capable of making realistic drawings of things for which she lacked concepts. Nadia was "diagnosed with what is now known as autism spectrum disorder" and had a vocabulary "limited to about 10 words" (2016: 74). Her drawings could not be considered "belief-dependent feature tracking." According to Walton's distinction, Lopes argues, Nadia's drawings should then be "transparent," like photographs. Hopkins' position, which conceives of photography as "*mind*-independent" rather than "*belief*-independent," would avoid Lopes' line of argument. Even if *belief-independent*, Nadia's drawings would still be produced via a chain of *mind-dependent* sufficient causes. This distinction allows them to be differentiated from photography. Lopes dismisses the importance of this distinction, claiming in a footnote that the difference between belief-(in)dependent and mind-(in)dependent "is not a difference that explains the epistemic differences between photographs and drawings" (2016: 161). However, I must admit that I do not clearly understand how Lopes' assertion can be sustained. Yet, in a 2010 article, Hopkins accepted the idea that "traditional photography, in contrast, involves a causal chain free from the influence of people's *beliefs* and experiences" (2010: 7, emphasis added).

This does not mean that authentic photography is completely independent of people's actions and intentions. There are intentional interventions that preclude imprinting and interventions that do not. If the relations causing the image are based on a mind-independent chain of sufficient causes, the photograph is authentic, even if many factors generating these chains are intentional actions. For example, choosing the scene to photograph, determining the equipment and lens, loading the film, and pressing the shutter are intentional actions. But once the shutter is pressed, the imprinting of the image still occurs, as "agency is exhausted in causing the relevant causal factors to be as they are" (2015: 331). Hopkins describes these interventions that do not preclude imprinting as a "second-order" form of control. However, removing or including elements in the image using darkroom techniques or combining elements in photomontage are mind-dependent processes that would change the pictorial content. In these cases, her intervention "is first-order control: altering content by herself entering the chain of sufficient causes" (2015: 333).

Hopkins is careful not to conflate *authenticity* and *accuracy*.[3] An image can result from mind-independent imprinting and still have distortions. As the author states, "photography's being authentic is no guarantee that the camera will not 'lie'" (2015: 333).

The central question Hopkins explores is to what extent a photographer can use the resources of authentic photography to communicate thoughts in an artistically interesting way. This question aims to determine the communicative limits of photography as a medium for artistically relevant thoughts. Here, the content of thought is communicated through *what* the photograph depicts and *how* it depicts it. Communication occurs when the spectator recognizes that the thought was intended by the photographer and is meant to be understood as such (we will explore this *Gricean* aspect of communication later).[4] Hopkins concludes that the resources available to photographers are far more limited in authentic photographs than those available to painters. But what resources are available to a photographer for communicating thoughts through authentic photography? According to Hopkins, these resources are "content, vehicle properties, relations between those two, and means of production" (2015: 334). These would be the means of photographic communication.

The content of a photograph is what it depicts—the objects visually represented and their properties. In Chap. 2, we will focus on what determines the pictorial content of an image and how contextual elements influence that content. For the purposes of understanding Hopkins' theory of artistic communication, it is important to note that he adopts a broad understanding of pictorial content, accepting that communication through content includes elements beyond the representation of objects and properties. According to him, it is also possible to communicate thoughts through the angle at which things are presented and how they are spatially arranged

---

[3] Similar to Walton, who distinguishes between *transparency* and *accuracy* (see 2008a, b: 33).

[4] Hopkins uses the term "audience" to describe those to whom the photograph is used to communicate. However, the auditory connotation of "audience" seems to introduce unnecessary complications. I will use the term "spectator," whose etymology is related to the Latin verb for seeing.

within the frame of the picture. In this case, the "geometrical structure" of the pictorial content serves as a "device for communicating thought" (2015: 336).

The vehicle properties are those possessed by the image as a material object, some of which are irrelevant to picture appreciation (e.g., the weight of a picture). However, some of these properties are artistically relevant to picture appreciation and are "resources to be exploited in making representational art" (2015: 334). When these properties possessed by the photograph are mobilized in determining the content, there is what the author calls the "interplay" between content and vehicle properties (2015: 330).

He distinguishes vehicle properties into three groups. The first group is *content-determining properties*, constituting what is commonly called the "design" of the image ("lines that compose the picture, what colors lie where on its surface, the boundaries between its colored areas, and the like" (2015: 337)). These are the properties by virtue of which the image has a certain pictorial content. In more technical terms, they are the literal properties of the image that are *syntactically relevant properties* (Kulvicki 2006: 30–31).

Additionally, vehicle properties include *content-neutral properties*, which he separates into two groups: *local* content-neutral properties and *global* content-neutral properties. An example of *local* content-neutral properties would be brushstrokes in a painting (which can vary in thickness and shape in different parts of an image), while *global* content-neutral properties might include the uniform surface of a photograph that can have a gloss or matte finish.[5]

The fourth resource available for a photographer to communicate thoughts is to explore the *means of production*. According to Hopkins, when a spectator is aware that a certain means of production was used by the photographer, the photographer can use that knowledge as a means of communication. The photographer can then explore peculiarities of the image production method to communicate thoughts.[6]

We can summarize the elements that a photographer has at their disposal, according to Hopkins, to communicate their thoughts as follows:

[5] One of the problems raised by Hopkins's distinction is how to interpret local content-neutral properties in photographic terms. In painting, it is plausible to think that the same scene can be represented using different kinds of brushstrokes, varying, for instance, in thickness. But what would be the photographic counterpart of this? This difficulty will only be addressed in Sect. 1.2, where we examine Paloma Atencia-Linares's (2018a) critique of Hopkins. One of the reasons for the communicational poverty of photography, according to Hopkins, is that the control over local content-neutral properties in photography is virtually nonexistent. As we shall see, Atencia-Linares argues that Hopkins is mistaken on this point, and that such control is greater than he initially assumes.

[6] This is a point that Hopkins addresses only briefly. Hopkins does not provide any explicit examples of how means of production could be used for communication. Later, I will argue that different photographic practices have distinct constitutive and regulative rules, which allow spectators familiar with the practice to have assumptions about the mode of production and the elements available to the photographer within the practice to vary this production. This common knowledge about production will play an important role in photographic communication.

1. Content: What the image depicts, its visual properties, and how it depicts them (including geometry).
2. Means of Production: The way the image is produced.
3. Vehicle Properties:
    (a) Content-Determining Properties: Literal properties that have a representational function.
    (b) Local Content-Neutral Properties: Properties without a direct representational function related to content that can vary within the image.
    (c) Global Content-Neutral Properties: Properties without a direct representational function related to content that are uniform and constant in the image.

According to Hopkins, the central element of artistic communication is the exploration of the relationship between *vehicle properties* and *content*, which he calls "interplay" (2015: 330). Hopkins' central thesis is that even though authentic photography has all these means at its disposal, the elements it has to achieve this interplay are limited compared to painting.

Hopkins' argument in support of this thesis has five premises and starts from the following reflection. For Hopkins, vehicle properties will play a communicative role only if they are "under the artist's control in a way that appreciators can detect" (2015: 336). It is at this point in his analysis that the foundation of his theory of artistic photographic communication in Paul Grice's theory of communication (1957, 1989) becomes evident (we will explore this point in detail in the following sections, Sects. 1.2 and 1.3). Based on Grice, Hopkins argues that communication is not just about making someone grasp a certain thought, but the interlocutor "must also recognize your intention: they must grasp the thought by recognizing that it is the thought you want them to grasp" (2015: 336). Jerry Fodor refers to this element of Gricean communication as a *reflexive intention*, "the intention that the effect be brought about by the [spectator]'s recognition that the [picture] is intended to bring it about" (2012: 63).[7] Hopkins' acceptance of the Gricean thesis of reflexive intention leads him to the idea that, for communication to occur, the producer of the photograph must make their intention to communicate a certain thought manifest, and the spectator must grasp how this intention is manifested, recognizing that the thought was intended. But for this to happen, the means the photographer has at their disposal to manifest their thought must be elements that the photographer can *intentionally and manifestly manipulate*; otherwise, appreciators would not be able to detect the intentional aspect in the use of this element. It is due to this reflexive aspect that, in the case of artistic photography, one can ask why the photographer chose those specific features.

These ideas lead Hopkins to his first premise:

1. If interplay is to play a role in the communication of thought, and thus in communicative representational art, it must lie under the artist's control in a way that appreciators can detect (2015: 336).

[7] Here, I adapt Fodor's ideas to the context of pictures.

The second step of his argument is to relate interplay to the three types of vehicle properties. According to the author, in his second premise:

2. To have suitable control of interplay, the artist must control vehicle properties independently of content. The candidates are content-determining, local content-neutral, and global content-neutral properties (2015: 337).

Hopkins offers an important argument to support his idea that the vehicle properties primarily available to the photographer for communicating thoughts will be *content-neutral* properties. The way I interpret his argument is as follows. Different properties that determine content could effectively portray the exact same scene with equal precision—such as in the case of painting, which can depict the same scene in oil or acrylic. It would be within the photographer's reach (albeit in a very limited way) to vary content-determining properties without altering the content. The photographer can generate the colors and shapes that constitute the design of a picture by different means. But because these properties determine the content, the spectator would be unable to discern, without additional information, whether these vehicle properties were intentionally used by the artist or whether they were used "because she wanted the picture to bear that content" (2015: 337). The use of content-determining properties for communication would thus be unable to fulfill the Gricean requirement of reflexive intention in communication. The spectator would always be left wondering whether the use of these properties was intentional or a requirement for the image to depict what constitutes its content.

However, it could be the case that the choice of certain content-determining properties is made intentionally, with some communicative purpose, even if the spectator could not determine this by merely observing the image. In such a scenario, the spectator would need *background knowledge* of the work and an understanding of the artist's intentions regarding it. However, in this case, Hopkins argues (in the important footnote 6 (2015: 337)), it would not be the resources of that specific art form doing the communicative work, but rather the background knowledge. If photographic communication always depended on background knowledge, argues Hopkins, there would be something fundamentally flawed with that art form as a communicative representational art. (I will return to this point in Sect. 1.2).

Moreover, the possibility of varying content-determining properties is always very limited because, depending on how the photographer varies these properties, it could change the image's content such that it would no longer be produced by "a mind-independent chain of sufficient causes." In other words, depending on the level of manipulation of content-determining properties, the image's content might no longer result from *imprinting*, and the photograph would cease to be *authentic*.

In the case of painting, the artist has extensive possibilities to use content-determining properties without altering the content. For an image to depict a certain scene, it just needs to be composed of a pattern of the right shape, even if it can be done "in oils, in ink, with chalk on slate, or carving into wood, marking boundaries with the edges of colored areas or by drawing lines" (2015: 338), and so on. Additionally, in painting, the artist can exercise broad control over both forms of *content-neutral properties*. They can alter local properties (such as brushstrokes) without affecting the content, or they can alter global properties (such as the size of the canvas). These considerations lead Hopkins to his third premise:

3. Painting offers the artist this control in respect of all three candidate properties (2015: 338).

In the case of photography, the use of content-determining properties as a vehicle for communication is unpromising because the photographer is caught between a rock and a hard place. On the one hand, the use of these properties would be difficult for the spectator to detect as an intentional act by the artist. On the other hand, if the artist manipulates these properties to an evident extent, it could alter the content, causing the photograph to lose its authenticity. Therefore, photography is left with only two candidate properties for communication: *local* and *global* content-neutral properties.

Certainly, the photographer can significantly alter global properties such as the image's size and the gloss or matte quality of its printing. Hopkins considers the example of Shomei Tomatsu's photograph "Beer Bottle after the Atomic Bombing" (Museum of Modern Art, New York), which depicts a bottle twisted and charred by the destructive power of the atomic bomb (2015: 341–342). According to Hopkins, the way the artist prints the photograph in a high gloss finish has the power to establish (or contribute to) an analogy between the bottle and scarred flesh, as the image has the reflectance of burned skin. This analogy that would be lost or weakened if the photograph were printed on matte paper. In this photograph, there is an important *interplay* between the global content-neutral property and what the "work communicates, via the analogy between the bottle and the flesh, about the effects of the bombing on its victims" (2015: 339). (This point will be very important for the subsequent analyses in this book, concerning how the scope of my analyses differs from Hopkins'. I will return to it in Sects. 1.2 and 1.6).

In this example from Hopkins, we can see how he introduces a distinction that will be fundamental to this book. There is a crucial difference between the *content* of a picture (in the case of Tomatsu's work, a twisted and charred bottle) and what the work *communicates* (something about "the effects of the bombing on its victims"). One of the purposes of this book is to explore this distinction and propose ways to understand how we transition from the pictorial content to what the use of the photograph communicates. This involves understanding what the user of the picture intends to communicate by ostensibly using a particular image with a certain pictorial content in a specific context. Hopkins, therefore, has the merit of precisely identifying a very important distinction between two levels of photographic communication: what the image depicts (its content) and what the intentional use of the image communicates.[8]

I believe Hopkins correctly understands that photographic communication is not merely about depicting something in a certain way. For Hopkins, it also involves

[8]This will later allow us to separate pictorial content from the *speech act* performed through the use of the picture.

depicting something using specific constitutive elements of photography, employing certain means of production, with the intention of communicating something to an interlocutor in a specific context (usually within an already established and consolidated social photographic practice). But (as we will see in subsequent sections) I will diverge from three fundamental elements of Hopkins' approach: *i)* the excessively restrictive aspects of Grice's communication theory, *ii)* the centrality of the notion of authentic photography, and *iii)* the understanding of communication as limited to artistic use. For now, let's return to Hopkins' argument.

The decisive point of Hopkins' thesis on the communicative poverty of photography as representational art is that authentic photography offers little beyond the use of global content-neutral properties for interplay with the content. Regarding local properties, photography could alter colors and shapes, but this would have to be done in an extremely limited way because, if the variations were significant, they would be content-determining properties. According to Hopkins, photography lacks brushwork, impasto, or incision, and variations in image texture would be global properties (points criticized by Atencia-Linares (2018a), as we will see in Sect. 1.2).

Hopkins' partial conclusion is as follows:

> [T]he upshot is that in photography every locally varying property of the vehicle is absorbed into the task of determining content. The authentic photograph seems to aspire to efface itself, wherever possible lacking any detectable properties other than those content requires. (2015: 339).

In this perspective, for authentic photography to be a communicative representational art, its primary (or almost sole) path would be to explore printing techniques that allow the artist to control content-neutral properties, generally limited to *global* content-neutral properties.

Hopkins points out a serious danger if this becomes the primary mode of artistic communication via authentic photographs. We might be led to fracture the photographic practice into two activities: the imprinted image and the printing processes. This occurs when communication relies on what the author somewhat pejoratively calls "flamboyant printing" (2015: 342). For photography to be authentic, imprinting must be assured, and communication would then occur through the way the artist makes their thought evident to the spectator via the printing processes. According to Hopkins, Tomatsu's merit lies in using the printing processes without creating a significant fracture between the imprinted image and the printing processes, by using the global content-neutral property "in an understated way" (2015: 342). However, for authentic photography to remain a unified activity (without the aforementioned fracture), "even cases such as Tomatsu's must remain the exception" (2015: 339).

These considerations lead us to the author's fourth premise:

4. Authentic photography offers the artist only limited scope for controlling interplay—i.e., with respect to global content-neutral properties alone (2015: 343).

From Hopkins' perspective, the limitation of authentic photography as a communicative representational art arises from the central role of interplay. In representational art, the content of the thought being communicated cannot be separated from the manner in which it is communicated. This inseparability is essential to the aesthetic appreciation of representational art *as* representational art. This idea is encapsulated in his fifth and final premise:

5. The exploitation of interplay is central to the full development of communicative representational art (2015: 343).

Based on the path outlined by the five previous premises, the fate of authentic photography is sealed, leading to the conclusion:

> (C) Authentic photography, unlike painting, is not able to develop to the full as a communicative representational art (2015: 344).

1.2 Criticisms of Hopkins' Theory

In this section, I will examine some of the criticisms directed at Hopkins, several of which are proposed by the author himself. I will also explore some points of divergence between Hopkins' position and the photographic communication theory I intend to develop throughout this book. Special attention will be given to the critiques put forward by Paloma Atencia-Linares (2018a). One of Atencia-Linares' criticisms explores how Hopkins bases his understanding of communication on Paul Grice's theory. This point will have important consequences for the subsequent analyses I will develop. We will note in Sect. 1.3 how abandoning one of Grice's theses will allow for a broader understanding of photographic communication.

A primary critique that Hopkins addresses is that his analysis centers photographic communication on the idea of *interplay* between vehicle properties and content and does not consider the possibility of communication through what he classifies as "non-depictive content". Hopkins presents two cases of communication that use non-depictive content. The first is when the photographer depicts one thing with the intention of symbolizing another—for example, "the dove and peace" (2015: 345). In Sect. 2.3, we will see that this type of pictorial denotation by conventional symbol can be understood as a form of *iconographic reference*.[9] The second case is when a photograph might depict a particular to exemplify a *type*. In our later analyses of pictorial denotation, I will call this case "pictorial generalization". Hopkins illustrates this type of communication through "Dorothea Lange's famous images of suffering in the Great Depression" (2015: 345). One of Lange's famous photographs, titled "Migrant Mother" (1936) (Museum of Modern Art, New York), portrays the suffering and resilience of Florence Owens Thompson. However, according to Hopkins, Lange's communicative intention is not just to depict the suffering and resilience of one person but to *exemplify*, through Thompson, the suffering and resilience of American women and people during the Great Depression. There is a *non-pictorial* process of generalization in which one might depict a

[9] See Lopes (2004) and Kulvicki (2006).

particular to exemplify a type. (This generalization is present in pictorial dictionaries using photographs and in the use of photographs in shopping catalogs, among other uses).

Although Hopkins acknowledges that art can use these forms of non-depictive communication, he argues that each art has a form of content that is basic to that art. In the case of pictorial representational arts, this basic content is *depictive*. Thus, even though there is the possibility of nonbasic communication in a non-depictive way (through iconographic reference and generalization), what is constitutive of photographic art as representational art is the basic communication in a depictive way.[10] Therefore, the exploration of the vehicles that an art form has for basic depictive communication is central to the appreciation of that art in its specificity. The exploration of nonbasic content is "incidental to the main plot" (2015: 346).

Another critique that can be addressed to Hopkins, which the author explores, is the fact that his conclusion applies exclusively to authentic photography. However, if we consider the ordinary uses of photography, such as in our daily communicative practices, which almost entirely occur through digital photographs, very few images would fit what the author classifies as authentic photography. And if we consider the use of photography in art, especially in contemporary (postmodernist) art, very few artistically relevant current photographic images would be authentic. *Why then worry about a conclusion that pertains to such a restricted and insignificant portion of our ordinary and artistic photographic practices*?

According to Hopkins, the problem his argument identifies with authentic photography can be extended to include non-authentic photographs, as many non-authentic photographs are made to resemble authentic ones. It is common, for example, to strive to make digitally manipulated images look like photographs that have not been manipulated. Thus, photography is again caught between a rock and a hard place; even though abandoning authenticity expands control over the use of vehicle properties and how they determine content. If the photographer exercises this control in an evident manner, the image will no longer look like an authentic photograph. But if this manipulation occurs subtly (so that it appears authentic), the spectator will not pick up on the artist's intentions, and the image will struggle to meet the Gricean requirement for communication. Without background knowledge, this intentional use would be difficult to detect. For communication to occur, the artist must make their intention manifest, and the spectator must recognize that it was intended by the artist. Non-authentic photographs that resemble authentic ones would face the same communicative limitations as authentic photography when it comes to manifesting and recognizing the intentional use of vehicle properties and the interplay between these properties and the content.

Another line of argument against Hopkins, which he considers, is that the centrality he grants to the idea of interplay between vehicle properties and content is tied to an outdated notion of art, typical of modernism. Although Hopkins does not

[10] This position is similar to that of Dominic Lopes, who emphasized the importance of "basic pictures," where the pictorial content is recognized by "suitable perceivers." According to him, "non-basic pictures always involve basic depiction at some level" (2004: 172).

formulate this point based on Clement Greenberg (1959), the problem he explores can be traced back to Greenberg's characterization of modernist art. As Greenberg explains, in modernism, each art form must carry out its self-definition through its own means. In the quest for purity of self-definition, modernist painting becomes increasingly about paint over a flat surface—these vehicle properties being the defining properties of *painting as painting*. In this case, the artist's intention to carry out the self-definition of painting is manifested in how the vehicle properties become evident and how these properties determine the content. Thus, in modernist art, the interplay between vehicle properties and content becomes the main means of communicating artistically relevant thoughts. But why impose these modernist limitations on art in general and photographic art in particular?

Hopkins argues that his level of generality is distinct from the central concern driving modernism (2015: 347). The problem he identifies applies to representational art *as* representational art, whether it is limited to modernism or not. For each representational art form, we could inquire to what extent it can explore the resources it has to construct its basic forms of representation. In this case, we would have to evaluate the capacity of each specific art to make the artist's intentions manifest through its vehicle properties, in the way these vehicle properties determine the basic content—which, in the case of representational arts, would be pictorial. Hopkins' argument is that if we evaluate comparatively, in relation to other arts, to what extent authentic photography can explore the resources it has for its basic forms of representation, we would find that authentic photography (or that which appears authentic) is not in a "fortunate position" (2015: 347). The communicative poverty of photography would be evident, even without considering the internal logic of modernism.

Important criticisms of Hopkins are formulated by Paloma Atencia-Linares (2018a). After a detailed exposition of Hopkins' central argument, Atencia-Linares uses a series of examples to show that photography has more ways of controlling content-determining properties in the interplay than Hopkins assumes. One of Hopkins' ideas is that photography lacks something that plays the role in painting akin to the control of brushwork and strokes (in which the painter can vary the thickness, intensity, and speed of the stroke). Atencia-Linares is very successful in showing that this is not the case. In photography, she argues, the control of lighting in a portrait (through the use of backlights, softboxes, and diffusers) could "define or blur, sharpen or soften the lines that determine the depicted objects by illuminating the set in different ways" (2018a: 89). Thus, the silhouette of the person portrayed could be more strongly or weakly delineated, and different facial features could be made more prominent or subdued (such as skin imperfections, contours, and proportions). According to her, this shows that photography could "vary the content-determining vehicle properties (things such as the quality and types of lines and patterns of lights and shadows) without changing the content" (2018a: 90). This would alter the way the object is depicted in the image without changing the object that is the image's content. And none of these manipulations precludes photography from being authentic.

Atencia-Linares also argues against Hopkins, asserting that photography has more control over local content-neutral properties than Hopkins assumes. According to Hopkins, in photography, the control of local content-neutral properties is almost nonexistent. However, selectively controlling lighting within the scene can emphasize or de-emphasize different elements of the image locally. Moreover, the surface of the image does not necessarily have to be thought of as a global content-neutral property, as Hopkins assumes, since a photograph can produce or use a surface with different textures in different areas. In this case, the differences in the surface would be local vehicle properties. Regarding the speed of the brushstroke in painting, Atencia-Linares argues that there is a parallel with the use of long exposure in photography. Long exposure marks the slowness with which the photograph is captured, and this can create effects in different parts of the scene, making it a local vehicle property (2018a: 91). These considerations by Atencia-Linares show that, in practice, the photographer has more control over the interplay between vehicle properties and content than Hopkins assumes.

But the main contribution of Atencia-Linares, concerning the purposes of this book, is highlighting how the theory of photographic communication in Hopkins relies on Paul Grice's model of communication and that altering the model of communication could lead us beyond the restrictions Hopkins imposes on photography as communicative representational art. The acceptance of the Gricean model is explicit in Hopkins (2015: 336) (as well as in Walton (2008a: 39–40)). A fundamental theoretical element of Grice is the distinction between *natural* and *non-natural* meaning (1957). In natural meanings, the relationship that an expression has with what it means is based on a causal and non-intentional relationship—such as how dark clouds naturally mean rain. Thus, a key attribute of natural meaning is its factive nature: "*the fact that something naturally means X implies X*". Grice exemplifies the category of natural meanings with photography. Non-natural meaning, on the other hand, is not confined to such a factive relationship. The connection between an expression and its non-natural meaning is not merely causal. Grice illustrates non-natural meaning with the example: "those three rings on the bell (of the bus) mean that the bus is full" (1957: 377). His theory primarily seeks to elucidate the use of non-natural meaning.

The point to note is that, for Grice, the ostensive act of indicating a natural meaning (when proposing, for example, a photograph as a candidate for attention in an interlocutory context), even if it were a manifest act with the intention of communicating something, would not constitute genuine communication. Even if we recognize that by showing a certain photograph the user does so with a certain communicative intention, the fact that what is communicated could be inferred from the inspection of the photographic image itself (because the photograph *naturally means* it) would make the communicative intention unnecessary. The same applies to the ostensive use of natural meanings in general. Pointing to the stormy sky, intending to communicate to someone to take an umbrella, would not be genuine communication for Grice because this thought could be inferred from the stormy sky itself, making the intention to communicate it unnecessary. In other words, in the use of photographs (as in the ostensive use of natural meanings in general), it would not necessarily be through the recognition of communicative intention that the interlocutor would grasp the communicated thought, as they could infer this information directly from what something naturally means. (As we will see in the next section, this conception of Grice can be criticized).

Atencia-Linares correctly states that it is plausible to conjecture that Hopkins accepts these general ideas of Grice in how he articulates intentionality and photography. For Hopkins, merely showing a photograph intentionally would not be communication, even if the communicative intention were manifest by the photographer and recognized by the spectator. For there to be communication via photography, according to Hopkins, the elements the photographer uses in producing the image "must lie under the artist's control in a way that appreciators can detect" (as stated in his first premise). The ostensive use of photographs in communicative contexts would not be genuine communication because what an authentic photographic picture naturally means would be independent of the artist's control. (It would be independent in the sense of being the result of a mind-independent chain of sufficient causes, in which the artist has only second-order control). What an authentic photograph depicts through a mind-independent chain of sufficient causes would not be something someone could genuinely communicate.

From this perspective, the assumption driving Hopkins' reflection is the acceptance of Grice's general theory of communication, that for photography to constitute a genuine form of communication, the means the photographer has to communicate their thoughts through photography would be restricted to what they can intentionally control without altering the image's content. In other words, photographic communication would be restricted to the properties of the image that the photographer could use as non-natural meaning, whose alteration does not preclude it from being an authentic photograph. It is by accepting the Gricean thesis that Hopkins conceives interplay as the means through which thoughts could be communicated in an artistically relevant way. Given this restrictive scenario, it is not surprising that Hopkins concludes that photography is poor as a means of communication.

Atencia-Linares argues that "if Hopkins is following Grice's model, it seems that he needs to justify why he does so" (2018a: 95). She contends that photographic communication does not need to be restricted to Grice's communication theory. If we accept Sperber and Wilson's Relevance Theory (1995), which is a development of Grice's theory, showing photographs through ostensive behavior that is recognizably intentional could be full-fledged cases of communication. (This will be the path I follow in the upcoming chapters).

Although Paul Grice did not develop a philosophy of photography, his critique of communication through natural meanings using photography as an example provides an important starting point. In the following section, I will develop Grice's presumed stance on photography to identify which of his theses needs to be abandoned or revised to achieve a more extended understanding of photographic communication than what we find in Hopkins (building on the insight proposed by Atencia-Linares).

The theory of photographic communication that I will propose in the following chapters will significantly diverge from Hopkins' proposal. It is the way Hopkins develops his photographic version of Grice's theory that leads him to conceive communication as being based on the relationship between the *producer* of a photographic image and a *spectator*, considering which elements the producer has at their disposal to express their communicative intention without precluding the photography from being authentic. My aim will be to broaden the scope of photographic communication to include cases where we use photographs, whether authentic or not, with a communicative intention, regardless of whether or not we are the image's producers. Much of ordinary communication via photographs involves using images we did not produce. This shift will move the focus of philosophical analysis from the *producer-spectator* perspective to the *user-spectator* perspective.[11] This does not mean that the producer-spectator perspective is irrelevant or unimportant in photographic communication, but we can think of it as a particular case of the user-spectator perspective, where the user is also the producer of the image. Additionally, my analysis will not be limited to understanding communication within the context of art, or more precisely, to what Hopkins calls representational art. I aim to understand communication through photographs in its ordinary form, whether artistic or not.

These changes will alter the temporal perspective of communication, as I intend to demonstrate in the following paragraphs. For there to be photographic communication in Hopkins' view, the requirements would be:

- The producer must have the intention to communicate a certain thought *at the moment* of image production (or a moment prior to production);
- This intention must be made manifest through the interplay between vehicle properties and content;
- The spectator must recognize that this interplay was done intentionally;
- Communication occurs when the spectator recognizes that the thought was intended by the producer of the photograph and that it was meant for the spectator to understand it in virtue of its being intended.

One danger of this perspective is that it could lead to a *mythology of the artistic creation process*. From Hopkins' perspective, the artist must possess a propositional representational content they want to communicate *at the moment* of image production (or prior to it). They then produce the image intending to make this thought manifest, expecting that the spectator will recognize the thought the artist aimed to communicate. However, in my practice as a photographic artist (which began in the late 1990s) and my work as a consultant and art critic with dozens of photographers

[11] The user does not need to be a person, an individual, but can also be a legal entity, such as a corporation, a company, a store, etc. In advertising, for example, when a store uses a photograph as an advertisement, there is a *directive intention* (in the sense that it intends for the viewer to perform the action of buying the product) and a *commissive speech act* (in the sense that the picture determines what the company commits to delivering if the product is purchased). In this case, the speech acts are performed by a legal entity, a company.

(many of them renowned artists) and numerous students, artistic production rarely conforms to the model envisioned by Hopkins. Often, the understanding of what the artist intends to communicate through their photographs is something achieved many years after the image's production, depending on interlocutory and maieutic processes with curators and editors. After this long process, when we hear the photographer's account of their past work, we have the impression that this representational content was present at the moment of image production. But this is not always the case. It is also not uncommon for an artist to revise their view about the communicative intention of their work when presented with a more precise or compelling interpretation than the one they originally held. Artists may also begin to shape their own narratives about their communicative intentions based on third-party interpretations—interpretations that themselves may evolve over time.

One might argue, in defense of Hopkins' position, that the artist already had this communicative intention at an unconscious level and merely needed help to articulate it in the appropriate words. One might also argue that the possession of this prior communicative intention, as an articulated representational thought, is present in *true artists*—who we might call "geniuses." But here, our mythology of artistic creation would be even more divorced from ordinary artistic production practices, as we would have to assume that a significant portion of people engaged in artistic production are not genuinely artists or that their productions are driven by unconscious *representational* thoughts.

Even though these considerations about the artistic creation process are not a refutation of Hopkins' position (since it is based on empirical observations related to artistic production practices, which can vary from place to place), I believe it is important to have a theory of the communicative uses of photography that aligns with how artistic creation processes occur in practice.

One way to avoid the necessity of this *mythology* of artistic creation is to temporally shift the central axis of photographic communication from the moment of production to the moment of the picture's use, which could differ from one use to another.[12] The understanding of photographic communication need not be centered on the production of the image but could be conceived as centered on the contextualized act of proposing a certain photograph as a candidate for attention in an interlocutory context (with the act of ostensively using the image with a communicative intention being described as a "photographic utterance"). This would explain how what a photograph communicates can change over time and vary from context to context and from user to user. In some cases, but not necessarily in all, there could

[12] What I propose here is very similar to what Daisy Dixon argues in her doctoral thesis on pictorial speech acts as shifting speech acts (2018). The difference lies in scope. Her analysis concerns images in general, whereas mine is more narrowly focused on communication through photographs, acknowledging that certain features of photography uniquely shape communicative possibilities. At the same time, my scope is broader in that I am not concerned solely with communication within curatorial or artistic contexts, but rather with photographic communication in ordinary settings. That said, it is important to acknowledge the insight found in Dixon's work and the quality of her thesis.

be alignment between the communicative intention of the image's use and the intention present (if present) at its production. However, this could diverge, and the image could be produced without the artist grasping the future communicative purpose for which the image will be used by them or others.

Hopkins comes close to this perspective by accepting that background knowledge of the work can be important in determining what a photograph communicates. However, he argues that for an art form to be a communicative representational art, communication must occur independently of background knowledge because, as seen earlier, if communication always depended on this type of knowledge, there would be something fundamentally wrong with that *art form* as a communicative representational art since "the art form's own resources are not doing the communicative work" (2015: 337). My goal is not limited to understanding photography as a communicative representational *art* but rather to understanding how we ordinarily use photographs for communication. Thus, it will not be a problem to assume that communication occurs without the communicative work being done exclusively by the photograph's own resources. Background knowledge is constitutive of any context of use and not a contingent addition to contextualized communication. Contextualized use always occurs within a background knowledge shared (to a greater or lesser extent) by the user and the spectator. Hopkins' artistic communication would be just one among many modes of photographic communication (though a rather restrictive one).

To illustrate the importance of context and the user in determining what a photograph communicates (which I will later parallel with the theory of speech acts, calling it the "photographic illocutionary act"), let us revisit an example from Hopkins. According to Hopkins, in Shomei Tomatsu's photograph titled "Beer Bottle after the Atomic Bombing," the way the artist prints the photograph in a high gloss finish has the power to establish an analogy between the bottle and scarred flesh. In this case, there is a subtle use of a global content-neutral property to determine what the "work communicates, via the analogy between the bottle and the flesh, about the effects of the bombing on its victims" (2015: 339). This is an example of successful photographic communication that avoids the problems of the rupture of "flamboyant printing."

But we could use a method *à la* Arthur Danto (1981) to highlight an important communicative underdetermination of photographs, which will require us to move beyond Hopkins' scope of analysis. Shomei Tomatsu's photograph depicts a twisted and burned bottle. Suppose now that this same picture (identical in all its visual and vehicle properties—and it is important to remember that photographs usually do not bring with them their causal histories) were the work of a young American artist who frequently depicts the nihilism of the American way of life in the interior of the United States. Suppose the young artist titles the work "Sunday Barbecue." Although we could still accept that the use of the high gloss finish could evoke the analogy between the bottle and scarred flesh, what the work communicates is no longer the effects of the Atomic Bombing on its victims. The analogy between the bottle and scarred flesh could still play an important communicative role, but it would communicate something about American eating habits and the American culture's propensity for the destruction of life.

It is because we know that it is a Japanese photographer depicting the aftermath of the Atomic Bombing and that the context in which the image is produced and used is that of highbrow photojournalism (with artistic aspirations) that the image communicates what it does (something about the effects of the bombing on its victims). If the picture were produced by a young Texan depicting the aftermath of a Sunday barbecue and used within an artistic practice typical of contemporary art, what the image communicates would be very different. The shift from the photojournalistic art context to contemporary art would also impact our expectations about the production method and equipment used. Photojournalism has constitutive rules that contemporary art does not. (This will be the focus of our analysis in Chap. 4). This would change the degree of epistemic confidence we have about what the image depicts. (Understanding how this impacts what the photograph communicates will further develop Hopkins' idea that the means of production play a crucial role in communication). In summary, what a photograph communicates is not determined solely by the content and the interplay of vehicle properties. The same image would communicate different things if produced or used by different people in different contexts and times. This allows us to conclude that photographs, when disconnected from context and use, are *communicatively underdetermined*.[13] Moreover, as David Novitz argues, an image is not always used for the purpose for which it was produced (1977: 67–85). One thing is the production of the image, and another is its use.

But Hopkins does not need to disagree with this idea. If communicative underdetermination is constitutive of photography, his position is even stronger, as he argues that photography is poor as a means of communication. We would be finding that it is even poorer than Hopkins assumed. This, however, does not seem to be a flaw of photography but something that extends and applies to pictorial representations in general. All of them would be equally underdetermined in terms of what they communicate—if disconnected from the context of use and without background knowledge.

We lose sight of the importance of context in communication when we assume it occurs within the ascetic world of museums and galleries—similar to scientific inquiry that relies on laboratory conditions (with one atmosphere of pressure, temperature of 25 °C, and no friction)—and with suitable spectators, where all share the necessary contextual knowledge for communication. *Understanding photographic communication is precisely understanding how underdetermination is overcome in the way we use photography as a means of communication in everyday life—a task central to this book.*

[13] In Sects. 2.2 and 2.3, I will relate this underdetermination to two other forms of photographic underdetermination: *pictorial underdetermination* (concerning the relationship between design and pictorial content) and *denotative underdetermination* (concerning the relationship between pictorial content and denotation). A fourth form of underdetermination concerns the fact that photographic pictures can be visually identical yet differ in their means of production. I will explore the fourth case in Chap. 4.

Furthermore, Hopkins does not deny the importance of background knowledge for artistic appreciation. He himself explores the problem of indiscernible artworks and argues that the aesthetic experience would differ in the case of identical works with different background knowledge (2005). The point I want to emphasize, by developing this same idea of Hopkins but in a photographic context, is that background knowledge not only alters aesthetic appreciation but is also necessary for communication and can change what the picture's use communicates. Since he does not explore this path, his theory tells us nothing about how to understand the role of context and the biography of the user/producer in photographic communication. It is precisely an approach that explores elements like these that I aim to propose through this book.

To expand the scope of analysis beyond Hopkins' Orthodox position, two fundamental steps are necessary. The first is the study of Grice's theory of communication, to identify what needs to be abandoned or revised for the use of photographs to constitute communication in a broader sense than just the interplay between vehicle properties and content. The second step is to find a way to broaden the understanding of photography beyond what Hopkins proposes as the essence of photography in his characterization of authentic photography.

Regarding this second step, the path I will explore to overcome the centrality granted to authentic photography is through understanding photography as a technology for producing differentiated surfaces by the action of light, as described in the New Theory (Sect. 1.5). This approach will give us the understanding that photography can play a role in line with what Grice calls *natural meaning* but also within *non-natural meaning*, creating a continuum between these extremes rather than constituting two separate and non-overlapping categories. A *discontinuous* approach is found in the understanding of photography as something that is essentially authentic or ideal (in the sense of Hopkins (2015) and Scruton (2008), respectively), such that a photograph is either authentic or not, ideal or not ideal. In practice, what we have in the use of photographs are pictures with varying degrees of epistemic reliability, stemming from the extent to which photographic social practices control the rules of image production. The idea of authentic photography would be a sublimation of how photography functions as a reliable informational system in rigorous epistemic practices, which have various constitutive rules for production. In some of these cases, we can make inferences about the world from the pictorial content and treat such images as factive. But this can occur to a greater or lesser extent, and there are photographic social practices whose function is not epistemic and accept that photographs can be fictional (non-factive) (as will be seen in Sect. 1.5). These ideas will be explored in this book through the development of an ontology of photographic social practices and the distinction between conventional and non-conventional contexts. In other words, my intention is not to refute or completely abandon the notions of authentic or ideal photography. Instead, I aim to propose a theory of photographic communication that applies to ordinary photographs, which, to varying degrees, approach or move away from these ideal forms. This theory will also seek to understand how this distancing or approximation occurs. Many elements must be in place for photographs to be used in an epistemically reliable manner.

1.3 Paul Grice and the Problem of Photography as Natural Meaning

In this section, my objective is to understand why, for Paul Grice, the act of ostensively showing a photograph with the intention of communicating something to someone would not constitute genuine communication. Achieving the idea of photographic communication through ostensive acts is essential to move beyond the perspective proposed by Hopkins (2015), which limits communication to the use of the interplay between vehicle properties and content (as analyzed in previous sections). This will also allow us to consider as communication those cases in which the image is used by someone other than its producer. Understanding why Grice does not consider the use of photographs as genuine communication requires understanding the significance he places on the interlocutor's recognition of the speaker's communicative intention.

Let's first examine the general outlines of Paul Grice's theory. Grice was one of the first to understand communication as a rational cooperative exercise, governed by principles and maxims, based on "a common purpose or set of purposes, or at least a mutually accepted direction" (1989: 26). The most general principle that governs communication is determined by Grice as:

> Cooperative Principle: "Make your conversational contribution such as is required, at the stage at which it occurs, by the accepted purpose or direction of the talk exchange in which you are engaged" (1989: 26).

In addition to the Cooperative Principle, Grice distinguishes four categories of maxims and submaxims that govern communication (1989: 26–27):

Maxim of Quality: "Try to make your contribution one that is true."

1. Do not say what you believe to be false.
2. Do not say that for which you lack adequate evidence.

Maxim of Quantity:

1. Make your contribution as informative as is required (for the current purposes of the exchange).
2. Do not make your contribution more informative than is required.

Maxim of Relation: "Be relevant."

Maxim of Manner:

1. Avoid obscurity of expression.
2. Avoid ambiguity.
3. Be brief (avoid unnecessary prolixity).
4. Be orderly.

A large part of communication occurs through *conversational implicatures*. Implicatures are inferences made during a conversation based on the assumption that the maxims are in operation or the evident breach of the use of the maxims. In a context where the maxim of Quality is in operation, there is a presupposition that the speaker says something true and based on adequate evidence. If someone says "it is raining," we can infer, based on the maxim of Quality, that the speaker has a belief based on adequate evidence that it is raining. If a speaker says "the car is running out of gas" and the interlocutor responds "there is a gas station at the corner," there is, based on the maxim of Relation, the implicature that the interlocutor believes the gas station is open; otherwise, the statement "there is a gas station at the corner" would not be relevant. However, if on a rainy day the speaker says "what a beautiful day," the evident falsehood of the statement allows for the implicature that she is being ironic, precisely by clearly violating the maxim of Quality.

But the determination of whether an implicature is present in the conversation depends on several elements beyond the maxims and the Cooperative Principle. According to Grice, to discover if an implicature is present, the listener will rely on the following data:

> (1) the conventional meaning of the words used, together with the identity of any references that may be involved; (2) the Cooperative Principle and its maxims; (3) the context, linguistic or otherwise, of the utterance; (4) other items of background knowledge; and (5) the fact (or supposed fact) that all relevant items falling under the previous headings are available to both participants and both participants know or assume this to be the case (1989: 31).

Based on these elements, the determination of the presence of a conversational implicature occurs as follows (1989: 31):

(i) The speaker says that p.
(ii) There is no reason to suppose that the speaker is not observing the maxims, or at least the Cooperative Principle.
(iii) The speaker could not be doing this unless they thought that q.
(iv) The speaker knows (and knows that I know that they know) that I can see that the supposition that they think that q is necessary.
(v) The speaker has done nothing to prevent me from thinking that q.
(vi) The speaker intends that I think, or at least is willing to allow me to think, that q.
(vii) Therefore, the speaker implicates that q.

A task we will undertake later (in Sect. 3.4) is to determine if communication via photographs is based on the Cooperative Principle and maxims. I will argue that it is, and this will allow us to determine how implicatures are present in communication via photographs (which can also be analyzed, in some cases, as *indirect speech acts*). However, before addressing the idea of photographic communication in a general sense, we must tackle a pressing problem: Grice's philosophy opposes the idea of communication via photographs, due to his understanding of the relationship between intention and communication and his distinction between natural and non-natural meaning.

In his famous 1957 paper, Grice uses photography as an example of *natural meaning*, distinct from genuine communication, which, in his view, hinges on the use of *non-natural meaning*. As mentioned earlier, a key attribute of natural meaning is its factive nature: "the fact that something naturally means X implies X". Consequently, if a photograph is indeed a photograph of a dinosaur, there must be a dinosaur that is meant by the photograph; otherwise, it wouldn't be a photograph of the dinosaur (Walton 2008a: 40). From Grice's standpoint, photographs are inherently factive. This, in a way, restricts the characterization of photography to natural meaning, similar to how Hopkins (2015) restricts his analyses to authentic photography and Scruton (2008) to ideal photography. A non-factive photograph would not be, in its authentic or ideal form, a proper photograph. In contrast, non-natural meaning is not confined to such a factive relationship. The connection between a sign and its non-natural meaning is not merely causal.

Regarding non-natural meaning, Grice distinguishes between what can be termed "speaker meaning" and "sentence meaning". Speaker meaning is what the speaker intends to mean, and sentence meaning is an abstract entity that the speaker can use on different occasions (see Borg et al. 2021). Speaker meaning and sentence meaning can diverge, as, for example, when a speaker utters the sentence "the door is open" intending to signify the command "close the door." According to Recanati, Grice's goal is "to analyze semantic notions such as sentence meaning or word meaning in terms of a pragmatic notion of communicative intention" (1986: 213). In other words, Grice aims to ground sentence meaning in speaker meaning—in what the speaker intends to mean.

As noted by Strawson (1964), Grice identifies three essential criteria for communication using non-natural meaning:

> S intends (i1) to produce by uttering x a certain response (r) in an audience A and intends (i2) that A shall recognize S's intention (i1) and intends (i3) that this recognition on the part of A of S's intention (i1) shall function *as A's reason*, or a part of his reason, for his response r. (1964: 446 – emphasis added).

These three intentions (i1, i2, and i3) are requirements that must be met for communication to occur genuinely, according to Grice. In contexts where what is communicated is a proposition, the speaker means, through the utterance of a sentence, a certain proposition to a certain interlocutor if and only if:

(i1) The speaker, by uttering the sentence, induces the belief in the interlocutor about the proposition;
(i2) The interlocutor recognizes the speaker's intention to induce in them the belief about the proposition;
(i3) The interlocutor's recognition that the speaker, by uttering the sentence, intends to induce in them the belief about the proposition is the reason the interlocutor forms the belief in the proposition.

Grice's use of (i3) leads to an extremely restrictive position where the recognition of the speaker's intention must be the reason or a necessary part of the reason that causes the interlocutor to form the belief that the speaker aims to communicate. Any scenario in which the interlocutor could form the belief without the recognition of the speaker's intention being necessary will not be understood as genuine communication—this is the central problem with natural meanings (and, consequently, with photography).

According to Grice, for communication to occur, we must not only intend to inform someone of something, but we must also intend for this intention to be recognized by the interlocutor, and this recognition must be the reason for the belief being produced in the interlocutor.[14] The speaker's intention that their intention be recognized by the interlocutor is called a "reflexive intention" by John Searle (1969), Bach and Harnish (1979), François Recanati (1986), Jerry Fodor (2012), among others. Communication is then thought of as something done overtly, where behavior makes it clear that the speaker is communicating, and it is necessary for the interlocutor to attribute to the speaker the intention that best explains their behavior. This distinguishes communication, which must be wholly overt, from other covert forms of information transmission. This is why Dan Sperber and Deirdre Wilson characterize Grice's theory of communication as "an exercise in mind-reading, involving the inferential attribution of intentions" (2002: 3).

Grice contends that photography fails to fulfill the third condition. When a photograph is used with a communicative intention, the message it conveys can be inferred directly from the image itself, as that is what the photograph *naturally* means. In such scenarios, the recognition of the speaker's intention (or, in this case, the photographer's intention) by the interlocutor is not deemed necessary. Grice uses an example to illustrate this point (1989: 218). I draw on his example. Consider a situation where Maria shows Claudia a photograph of Ana displaying excessive familiarity with Pedro, who is Claudia's husband. Maria's intention is to communicate to Claudia that Pedro is being unfaithful. According to Grice, this use of the photograph by Maria would not constitute genuine communication. This is because Claudia could form the belief about Pedro's infidelity simply by looking at the photograph, perhaps even if she stumbled upon the photograph. In Grice's view, Maria's communicative intention is not *necessary* to Claudia forming the belief.

According to Grice's criteria for genuine communication, the use of photographs satisfies only the first two requirements but fails with respect to the third:

(i1) By using the photograph, the picture user intends to produce a belief in the viewer.
(i2) The viewer recognizes the user's intention to produce such a belief in them.
(Fails at i3) However, the recognition that the user intends to produce the belief is not a necessary reason for the viewer to form the belief. The formation of the belief could occur independently of the user's intention, as the photograph allows for the belief to be obtained by naturally meaning it.

Grice contrasts this case with a situation in which Maria hands Claudia a drawing she made. For Grice, the use of a handmade picture in this case would genuinely communicate something to the viewer. The picture would embody the user's

[14] Some authors, such as Sperber and Wilson (1995), Strawson (1964), and Schiffer (1972), explore the possibility that Grice's theory of communication could lead to an infinite regress: "in principle, for any *n*th-order intention of this type, you need an $n + 1$th-order intention to the effect that the *n*th-order intention be recognized. In other words, you need an infinity of such intentions to explicate the intuitive notion of overtness along those lines" (Sperber and Wilson 1995: 31). This would lead, according to Sperber and Wilson, to the undesirable consequence in Grice's theory of making the speaker's meaning an idealization that is never achieved in real life.

intention in such a way that the formation of the belief in the viewer would depend entirely on the user's intention.[15]

The author draws a parallel between the case of photography and the biblical episode of the beheading of St. John the Baptist. After Herod divorced his wife Phasaelis and illegitimately married Herodias, he granted a wish to Salome, Herodias's daughter. Salome asks for the head of St. John the Baptist because he had publicly criticized Herod for divorcing his wife and marrying Herodias illegitimately. Herod grants Salome's wish and gives her the head of St. John the Baptist on a charger. The point explored by Grice is that when Herod presents Salome with the head of St. John the Baptist, he intends to produce in Salome the belief that her wish has been fulfilled. But Herod fulfills only the first two requirements of genuine communication. Just like in the case of the photograph, the recognition of Herod's intention is not necessary for Salome to form the belief that St. John the Baptist is dead. What causes her belief is the fact that his severed head is in front of her. The severed head *naturally* means that he is dead. Thus, Herod's intention is dispensable and does not form the basis of the communication of the belief in question.

If Grice's analysis of the use of natural meaning holds, it poses significant challenges to the development of a theory of photographic communication based on the idea that we could communicate through the ostensive act of showing photographs. The only option for photographic communication would then seem to be the one proposed by Hopkins, centered on the producer-spectator perspective and the use of the interplay between vehicle properties and content.

Several scholars, however, have found Grice's third requirement excessively restrictive (Neale 1992: 547–549; Schiffer 1972: 43–48; Sperber and Wilson 1995: 29, 50–54). According to Recanati, Grice would be correct in drawing a distinction between two cases (1986: 225). There are instances in which what is communicated depends solely on the speaker's intention. These encompass non-natural meaning and are central to linguistic communication. The other category consists of cases in which what is uttered provides additional evidence for what is intended by the speaker—such as in the case of the photograph and the head of St. John the Baptist. But, for Recanati, Grice offers no reason to deny that Herod communicates to Salome that St. John the Baptist is dead by showing her his severed head. Thus, there would be no grounds to confine communication to the first category. Recanati concludes that the notion that communication is never based on natural meaning "seems to me to be patently false" (1986: 225). For Recanati, communication through the use of natural meanings includes instances where: "I intend my utterance to provide the hearer with some evidence independent of the evidence provided by my recognized intention" (1986: 225–226). Thus, from Recanati's perspective, we could conceive of cases in which communication uses natural

[15] This example will be important later on (Sect. 1.5) because, by abandoning the centrality of authentic photography and considering photography in the framework of the New Theory (as proposed by Dominic Lopes (2016)), we will be able to extend photography to the category of non-natural meaning. In this way, a significant portion of photographs could have the same status as hand-made drawings.

meanings as genuine communication, as long as there is a communicative intention recognized by the interlocutor, even if what is uttered provides independent evidence of what is communicated.

Sperber and Wilson (1995, 2004) also depart from Grice's restrictive reading of the third requirement. It is not my intention to adopt all elements of Sperber and Wilson's "Relevance Theory" or to provide a detailed exposition of their theory, but rather to highlight where the main divergence from Grice lies. The idea of these authors is that their theory of communication comprises two layers: a *foundational layer of information* about any subject and a *second layer indicating that the first layer was intentionally manifested.* In this framework, the primary layer of information could be anything—including a photograph—while the communicative use hinges on the intentional manifestation of this information. So there are two types of intentions necessary to communication:

(a) *informative intention*: the intention to inform an audience of something.
(b) *communicative intention*: the intention to inform the audience of one's informative intention. (2004: 611)

For communication to succeed, both intentions must be fulfilled: the audience must grasp the content and recognize that the speaker intended to convey it. In this framework, to communicate is to ostensively point to a foundational layer of information, prompting the interlocutor to recognize that this information was intentionally highlighted (Sperber and Wilson 1995: 50). Communication is thus conceived as ostensive-inferential: the speaker performs an ostensive act that reveals a communicative intention, and the interlocutor infers from this act that the foundational content was intentionally conveyed. These two intentions jointly constitute the minimal conditions for successful communication.

The ostensive act of indicating the foundational layer of information can constitute genuine communication because, according to Sperber and Wilson, "an act of ostension carries a guarantee of relevance, and that this fact—which we will call the principle of relevance—makes manifest the intention behind the ostension" (1995: 50). Thus, any overt act of communication, in which there is an ostensive communicative act, implies its own relevance. In the case of photography, it would not be necessary for the intention to be manifested exclusively through the interplay between vehicle properties and content. The very act of intentionally proposing an photograph as a candidate for attention would be a way of making the communicative intention explicit. Faced with this image, the spectator could then ask what the user intends to communicate through the use of this photograph.

Sperber and Wilson (similar to Recanati) also acknowledge two types of cases, distinguished by whether the foundational information serves as *direct* or *indirect* evidence. For instance, a mother pointing to dark clouds as a direct indication of impending rain versus her stating "it will rain" as an indirect cue. Grice labeled direct evidence as natural meaning and indirect evidence as non-natural meaning, drawing a sharp distinction between these two extremes. Sperber and Wilson argue that there is a *continuum* between these two, where the foundational information can vary in its directness or indirectness. However, the critical factor for communication is the presence of a foundational layer and the intentional manifestation of this information, followed by the recognition that the first layer was intentionally targeted.

Thus, in Sperber and Wilson's development of Grice's theory, the use of photographs can constitute genuine communication if the fundamental information in the photograph is intentionally presented to the viewer by the speaker through an ostensive act—enabling the viewer to recognize that the photograph was deliberately targeted. From this perspective, there exists a continuum, ranging from instances where the photographic image serves as direct evidence of the speaker's intended message to cases where it functions as indirect evidence. Both cases would be considered communication, according to Sperber and Wilson, if the image is used in a manifestly intentional manner. The recognition by the interlocutor that the basic information (whether direct or indirect) is intentionally targeted would suffice for communication.

Richard Moore (2017), while acknowledging the validity in Grice's assertion that communication is not merely a causal process, criticizes the implication that naturally meant states cannot be used for communicative purposes. He argues that such a stipulation undesirably excludes animal communication from being considered genuine communication. Moore's approach also involves partially rejecting Grice's third requirement. He challenges Grice's interpretation of the episode of the beheading of St. John the Baptist. Moore proposes that Herod's use of the head could have served other communicative intentions beyond merely confirming the death. Herod could have employed it to convey additional messages, such as:

A. I have killed St. John the Baptist,
B. My word is my bond, or
C. This is how a good king rules (Moore 2017: 4).

The presence of St. John the Baptist's head on a charger does not unequivocally convey Herod's communicative intention. For instance, in scenario (A), the head could imply that John was murdered by someone else—thus, its use would not be communicating that "I [Herod] have killed St. John the Baptist." In cases (B) or (C), the head alone would not suffice to communicate the specific message intended. To understand the different communicative acts, Salome would need to discern Herod's communicative goals; the head itself would not naturally provide all the necessary information. If scenarios (A), (B), and (C) are considered genuine communication, then using the head to communicate that St. John the Baptist is dead should also be understood as genuine communication. Moore concludes that "the presentation of naturally meaningful states is, then, consistent with a system of communication in which hearers must attribute communicative intentions" (2017: 4). This perspective is crucial for Moore's argument, as it encompasses a significant portion of human and non-human communication that relies on the indication of natural meaning.

Applying Moore's analysis to the case of photography suggests that a single photograph can serve multiple communicative purposes (similar to the communicative underdetermination of photography discussed in the previous chapter). Grice's example, where a photograph of Ana showing excessive familiarity with Pedro implies his infidelity, overlooks the potential for varied communicative uses of the same image in different contexts. The determination of what information is being communicated hinges on understanding the user's communicative intention. Furthermore, contrary to Grice's assumption, a photograph of Ana showing excessive familiarity with Pedro does not naturally mean Pedro's infidelity; it merely portrays Ana's excessive familiarity with him. The same image could be used to communicate, for example, that a past animosity between Pedro and Ana has been overcome, or to convey many other messages. This suggests that Grice's interpretation of the photograph presupposes a communicative intention that extends beyond the visual information contained within the image itself and its causal connection to the depicted objects.

The criticisms proposed by Recanati (1986), Sperber and Wilson (1995), and Moore (2017) allow for a revision of Grice's position on the use of natural meaning as a means of communication. Grice's use of the third requirement leads to the idea that the use of photographs would not be genuine communication, as it would not be necessarily due to the recognition of the intention that the belief is formed in the interlocutor. To enable communication through the intentional use of photographs, we can move away from (i3) (as supported by Grice) and adopt a position similar to that proposed by Sperber and Wilson. According to this view, the use of photographs (and natural meaning in general) constitutes communication as long as there is an ostensively intentional use of a photographic image with the aim of conveying something to someone through the image. Photographic communication could then be constituted by two layers: the photograph as a foundational layer of information and a second layer indicating that the first layer was intentionally used. We could also accept, as Sperber and Wilson do, that the ostensive communicative act of using the photograph implies its own relevance.

This expansion of communication to include the ostensive use of photographs does not need to be taken as a refutation of Hopkins' position (addressed in the previous chapter). By setting aside the restriction imposed by the third requirement (i3), in light of the criticisms presented by the authors discussed in this section, we can broaden the notion of photographic communication to include the ostensive uses of photographs—without abandoning the idea that communication may also occur through the intentional use of the interplay between vehicle properties and content, and the recognition of the intention by the viewer, as Hopkins proposes.

This way, we can expand photographic communication beyond the *producer-spectator* cases (in Hopkins' perspective) to include *user-spectator* communication contexts, whether the user is the producer of the image or not. By including user-spectator communication contexts, we can shift the temporal perspective of communication (mentioned in the previous chapter) from the moment of producing the photographic image to the moment of the utterance act. It is at the moment when the photograph is ostensively and intentionally used in a user-spectator context that the use of the photograph can constitute an act of communication. Thus, the same photographic image could then communicate different things in different *utterance acts* if the communicative intention differs (or if there are variations in user and/or context). However, this does not refute Hopkins' position, as communication could still occur in the manner he describes. There is primarily a difference in scope here.

In Sect. 1.5, I will argue that it is possible to conceive of photography beyond Hopkins' notion of authentic photography, in such a way that photographs could be either factive or non-factive. Hopkins does not deny this possibility, but he does not provide elements for us to think about non-factive photographs (which would not be authentic). In cases where they are factive, photographs align with what Grice calls natural meaning. In cases where they are non-factive, photographs could be understood in accordance with what Grice calls non-natural meaning. I will accept that there is a continuum between these two extremes.

In subsequent chapters, I will explore Lopes's (2016: 45) idea that photography can be conceived as comprising various social practices of producing and using photographic images. The central idea I will propose is that different social practices have different *perceived functions*, which determine the *purpose* of photographs within these practices—what I will call, drawing on John Searle (1969, 1979), the "photographic illocutionary purpose". This will allow us to distinguish between conventional contexts of using photographic images (which will be practices with constitutive and regulative rules) and non-conventional contexts. Regarding conventional contexts, I will additionally argue that the ostensive act that manifests the user's communicative intention will not be crucial to communication, as the illocutionary purpose of the practice being intended will be an essential condition for participation within the practice. In other words, the mere presence of the image within the practice will be sufficient for the attribution of a communicative intention in line with the illocutionary purpose and function of the practice. In this case, we could say that the ostensive act is not necessary or is already presupposed. This approach enables a non-intentionalist perspective on photographic communication in conventional contexts. Instead of focusing on the ostensive intentional use of photographic images and the spectator's recognition of such intention, it emphasizes understanding the constitutive and regulative rules governing the practice and how the function of the practice determines the purpose of photographs. These rules determine how the illocutionary purpose of the practice is to be achieved. However, in non-conventional contexts (or those with a low degree of conventionality, since I will argue that there is a continuum between these extremes), the manifestation and apprehension of communicative intention will be fundamental for communication, as proposed by the revised Gricean theory.

1.4 The Orthodox Theory of Photography

In recent analytical literature, it has become commonplace to categorize philosophical approaches to photography into two major groups: the *Orthodox Theory* and the *New Theory*.[16] This distinction is significant for the purposes of this book, as it

[16] This segmentation of the history of the philosophy of photography, which categorizes authors into major groups, always risks exerting a certain violence by placing in the same group authors who often have significant differences among themselves. Any general categorization of photography risks overlooking important distinctions. By adopting this current terminology, it is not my intention to argue that the authors who fall under these two designations do not have significant

enables us to situate the previously discussed ideas of Robert Hopkins within the framework of the Orthodox Theory. The New Theory, on the other hand, emerges from the critique and deconstruction of the central ideas of the Orthodox Theory and proposes a significant shift in focus regarding the very object of philosophical analysis—as we will see in the next chapter. Understanding Hopkins's relationship to Orthodox theories and the critique of the New Theory will enable us to broaden our understanding of photography and, consequently, our comprehension of photographic communication. It will also allow us to redirect philosophical analysis towards the ordinary use of photography, thus avoiding the recurrent *sublimation* of Orthodox Theory, which distances itself from everyday photographic practices by prioritizing an analysis of its supposed *essence*.

In this section, I will outline the general principles of the Orthodox Theory, aiming to establish some points of convergence with Robert Hopkins. I will consider Roger Scruton (2008) and Kendall Walton (2008a) as paradigmatic examples of the Orthodox Theory.[17] In the next section, I will discuss the New Theory.

The philosophy of photography primarily focuses on four fundamental issues:

1. The ontological problem of determining what photography is.
2. The problem of distinguishing between photographs and handmade images.
3. The epistemological problem concerning the type of relationship we have with the world through photographs.
4. The problem of the status of photography as art.

According to Diarmuid Costello, the Orthodox Theory, "like other orthodoxies, [...] involves adherence to certain *idées fixes*" (2018a, 230). The *idées fixes* of the Orthodox Theory are integral to how its proponents address the four fundamental problems and articulate their answers. The foundational idea upon which the Orthodox Theory is built is the following general intuition about the ontology of photographic images:

> When contemporary philosophers maintain that photography is at bottom an automated process in which a mechanical apparatus produces images that depend causally and counterfactually on the scenes depicted in such a way as to bypass the beliefs and other mental states of the photographer, they build in a functionally equivalent *non-agential* explanation for the generation of photographs. (Costello 2018b: 230).

From this perspective, photographs are the result of using a mechanical apparatus that maps visual information in a purely causal manner (with a counterfactual

differences. For this reason, it would be more appropriate to use the plural terms "Orthodox Theories" and "New Theories." However, I will adhere to the current usage of the terminology (as in Abell (2018), Atencia-Linares (2018b), Costello (2018a, 2018b), and Lopes (2016, 2018)) by using the term in the singular, while remaining aware of the complexity that this terminology conceals.

[17] Orthodox theories are as old as the philosophy of photography itself and can be traced back to the thoughts of the creators of photographic processes in the nineteenth century. Moreover, as argued by Costello (2018a), these theories give philosophical expression to ideas and intuitions about the nature of photography that are part of our common sense.

dependence between the image and the depicted scene), free from the photographer's beliefs and mental states. Hopkins formulates this fundamental intuition of the Orthodox Theory as follows:

> Traditional photography [...] involves a causal chain free from the influence of people's beliefs and experiences in this way. [...] Once the shutter has been opened and the film exposed, the passage of information from one stage to another is essentially independent of how anyone thinks things were in the scene photographed. (2010: 7)

Authentic photography (described in his 2010 article as "traditional photography") represents, for the author, an *ideal* of photography, in which the objects, events, or arrays that served as input in the photographic process cause the pictorial content via a chain of mind-independent sufficient causes.

Based on the idea that photographs are intrinsically "causal," "mechanical," and "automatic," and free from the contamination of the photographer's beliefs and mental states, the Orthodox Theory adopts a *discontinuist* perspective regarding the relationship between photographs and handmade pictures.[18] In contrast to photographs, humanly mediated depictions are seen as "intentional," "hand-made," and "agential" (Costello 2018a: 9–10). The discontinuist thesis serves an ontological purpose, in the sense that photographs can then be defined by the specific difference that separates them from handmade images within the general category of pictures—allowing, in general terms, for an Aristotelian type of definition (based on genus and specific difference). Photographs and handmade images would then be species of pictures categorized into non-overlapping categories, although they are circumscribed within the general genus.

Scruton offers an important argument that consolidates this intuition underpinning the Orthodox Theory and establishes discontinuism:

> When the photographer sees the photographic plate, he may still wish to assert his control (...). He can proceed to paint things out or in, to touch up, alter, or pasticher as he pleases. But of course he has now become a painter, precisely through taking representation seriously. (Scruton 2008:156)

Any intervention by the photographer to "assert his control," thereby altering the content of the photograph, would turn the photographer into a "painter," breaking the purity of photography and distancing it from the ideal that constitutes its essence. Thus, the essence of photography is best embodied in unmanipulated, straight-out-of-camera photographs of the traditional Kodak type. Similar to Scruton, Kendall Walton argues that the paradigm of photography untainted by the photographer's beliefs and intentions (which he calls "transparent photography") "would seem to be not the work of professional photographers but casual snapshots and home movies made by doting parents and wide-eyed tourists with assists from Kodak" (2008a: 42).

---

[18] I am adapting to photography the *discontinuist* terminology used by Michaelian et al. (2020) to describe different positions in the philosophy of memory concerning the relationship between memory and imagination.

As previously seen (Sect. 1.1), in a more nuanced manner than Scruton, Hopkins (2015) accepts that the photographer can exert intentionality, but the photographer is limited to a "second order" form of control, in which their choices and interventions are exhausted in causing the relevant causal factors for the imprint to happen. Altering the content, which would be a form of first-order control, would result in inauthentic photographs.

The adoption of discontinuity between photographs and handmade pictures leads adherents of Orthodox Theory to a *sublimation* that shifts the focus of philosophical analysis from ordinary photography to a supposed essence of photography. If we analyze photographs as they are ordinarily produced and used, we will notice that there are countless ways in which the photographer can introduce their intentionality and beliefs, making photography "intentional," "hand-made," and "agential"—erasing the boundary erected by Orthodoxy between photographs and handmade images. It is not an exaggeration to say that the vast majority of contemporary photographic images (which are digitally created and manipulated by apps accessible to any smartphone user) would not fit the mechanical and causal purity ideal of Orthodox Theory. This *sublimation* stems from the idea that, even if our ordinary photographs do not fit the purity ideal proposed by Orthodox definition, photography in its essence would be pure. In Hopkins' terms, this purity is "central to photography's self-conception" (2015: 330). The Orthodox approach thus shifts the focus of philosophical inquiry from ordinary images—as we produce and use—to an alleged essence of the photographic image, which serves as a *normative ideal*. Thus, in Orthodox authors, philosophical inquiry often focuses on debating our intuitions about the essence of photography rather than understanding how ordinary photographs are used in our everyday practices. This leads to an almost insurmountable gap between philosophical analysis and our actual utilization of photography. This shift in philosophical analysis from ordinary use to an *ideal* of photography is what I will term "sublimation."

Based on the characterization of photography as "causal," "mechanical," and "automatic," the Orthodox Theory conceives the epistemological status of photographic images as privileged and distinct from that of handmade images. Photographs not only allow for *pictorial experience* (in which we see a scene depicted in the image), but this experience would also be *factive*. Photographs not only reflect the facts, but do so out of necessity, such that "if what is seen in the picture is that p, then p" (Hopkins 2010: 3).[19] This factivity is inherent in the Gricean notion of photography as natural meaning: the fact that something naturally means X implies X. Consequently, if a photograph is indeed a photograph of a dinosaur, there must be a dinosaur meant by the photograph—otherwise, it wouldn't be a photograph *of* the dinosaur, in the causal sense (Walton 2008a: 40). Through the thesis that photographs are factive, the discontinuity between photographs and handmade images

[19] The notion of "necessity" used by Hopkins is quite weak and should not be confused with a robust notion of logical necessity (as found, for example, in modal logic—something that is true in all possible worlds).

can be expanded, since handmade images are not factive. Even if handmade images correspond to facts, this correspondence does not occur out of necessity.

Robert Hopkins (2010) is one of the most ardent proponents of the thesis that photographs are factive and epistemically superior to handmade images. His defense of this thesis has two central claims. The first is the idea that the *goal* guiding the design and development of photographic equipment is to sustain *accurate* pictorial experience. The second claim is that, in virtue of this goal, photography is governed by *norms* that determine when things are working properly (Hopkins 2010: 8). Thus, when photography is operating properly, fulfilling the overarching goal of its design, it will provide an accurate pictorial experience of the depicted scene. So, when photography fulfills its point and purpose, it affords factive experience. Photographs could lie (be non-factive) only if the mechanism is not functioning as designed, which could occur due to malfunction or human interference.

The way Hopkins (2010) conceives the epistemic superiority of photography also leads to a form of the aforementioned *sublimation*. In practice, photographic systems will deliver inaccurate results, as it is common to sacrifice some accuracy for economy during the development and production of a system. This may be true for the vast majority of photographic systems. It could also be the case that no actual photographic system is *perfectly accurate* (whatever that could mean). A perfectly accurate photograph would be a normative ideal. In the face of these facts, his strategy is not to base the norms governing photographic systems on our *actual* norms, but on an *idealization* of them. Thus, what we have in practice, "while it does not strictly belong to a factive type, it is the next best thing" (Hopkins 2010: 20). Hence, the author concludes:

> Pictorial experience is factive when it is either the product of a system governed by ideal norms and where all has worked properly, or is the product of a system successfully governed by compromised versions of those ideal norms, and matches the experience the ideal system sustains, in terms of the putative facts seen in the picture (Hopkins 2010: 20).

Factivity as a distinctive epistemic trait will be found in the normative ideal of photography, exemplified by Hopkins's authentic photographs (Hopkins 2015), Scruton's ideal photographs (Scruton 2018), Grice's natural meaning (Grice 1957), and Walton's transparent photographs (Walton 2008a).

Supported by a causal theory of perception, photographs can be conceived by various Orthodox theorists as a form of *remote acquaintance*.[20] Photography would be an apparatus that allows us to bridge spatial and temporal gaps, connecting us directly to the depicted objects themselves; unlike handmade images, in which we only *fictionally* see something depicted. As Kendall Walton states: "I would subscribe to some variety of causal theory: to see something is to have visual experiences which are caused, in a certain manner, by what is seen" (Walton 2008a: 34). Thus, similar to mirrors, spectacles, and telescopes, photographs are aids to vision.

---

[20] I adapt here the terminology from the philosophy of memory to photography. Bertrand Russell conceives of remote memory as *remote acquaintance* (1984: 73). For more on memory as remote acquaintance in Russell, see Faria (2010).

Photography as remote acquaintance can be distinguished into two forms, with different levels of strength: *direct naïf realism* and *indirect realism*.[21] Roger Scruton, a proponent of the direct *naïf* realist variety, treats photography as a means of directly perceiving the object. He posits, "the camera, then, is being used not to represent something but to point to it. (...) The camera is not essential to that process: a gesturing finger would have served just as well" (Scruton 2008: 151). This perspective suggests that the camera acts merely as a finger, directing attention to the object itself rather than producing a representation of it. This leads to the complete devaluation of the aesthetic value of the photographic pictures. A beautiful photograph would merely be a cognitive remote access to an object that is beautiful. The aesthetic value, then, would always shift from the photograph to the photographed object.

On the other hand, Kendall Walton represents the *indirect realist* viewpoint. He acknowledges photographs as transparent representations, contrasting with the opacity inherent in handmade pictures. The discontinuity between photographs and handmade images arises from how photographs are considered to have a *belief-independent causal* connection to their subjects, whereas handmade pictures are seen as having a *belief-dependent causal* connection with their subjects (Walton 2008a). Except in rare cases (the famous *trompe-l'œil*), when we look at a picture, we do not have the illusion that the object itself is in front of us. What distinguishes the perception of objects from picture perception is that, in the case of pictures, we have, for Walton, a *twofold* experience—we see the content of the image (the depicted scene) and at the same time we experience the surface of the image. In the case of handmade images, we know we are looking at an image and, according to Walton, we use it as a *prop in a game of make-believe* (Walton 2008b). But in the case of photographs, due to the specific way they are causally connected to objects, although they are indeed representations, "my claim is that we see, quite literally, our dead relatives themselves when we look at photographs of them" (Walton 2008a, b: 22). Thus, for Walton, to see the object photographed in the picture is to see that object. Photography is a special type of picture in which the literal vision of the object is indirect, through the perception of a transparent representation.[22]

Although Hopkins accepts the factivity of photographic pictorial experience, he does not treat it as remote acquaintance. According to him, pictorial experience is not directed at objects (as supported by Walton) but at putative facts. Thus, due to

[21] I am adapting here to photography Denis Perrin's terminology from philosophy of memory (2012: 20–21).

[22] In Walton's perspective, there is an inherent tension in the perception of photographic pictures. This tension lies between the fictionality of direct vision (which applies to all images) and the literalness of indirect vision (specific to photography). When looking at a photograph, I can imagine, as part of a game of make-believe, that I am seeing the object *directly*—just as we do in the make-believe game of perceiving pictures in general. But this is fictional, in the sense that I do not have the illusion that the object itself is in front of me—I also see the surface of the image and know that I am looking at a picture. However, since it is a mechanically generated picture, in the case of photography, I literally (not fictionally) see the object *indirectly*. As Dretske explains, for Walton, vision in photography is indirect, as we see the object through the photograph (1984: 73).

the factivity, if someone sees in a photograph that *p*, then *p*. However, while for Walton, through a photograph of Churchill, we would see Churchill himself (even if indirectly, mediated by a representation), for Hopkins we see "that Churchill was fat" (2010: 4). But it would not make sense, according to Hopkins' position, to say that through a photograph of Churchill, we see the putative fact itself that "that Churchill was fat."[23]

The way the Orthodox Theory characterizes the essence of photography and distinguishes it from handmade pictures often leads Orthodox authors to a form of skepticism about the status of photography as art—although it is possible for Orthodox authors to adopt a *non-skeptical* position (as in the case of Walton, according to Costello (2018b)) or a form of *moderate skepticism* (as in Hopkins). But it is important to note that this skepticism is not a factual judgment about how we use photography, which is clearly accepted as art. The thesis that photography is not art (in Scruton's radical skepticism) or is only a limited form of art (in Hopkins's moderate skepticism) is articulated in relation to the supposed essence of photography. Although ordinary photographs are considered art, photography in its authentic or ideal form brings with it a degree of skepticism about the claims of photography as art. The strongest version of this sublimation is formulated by Costello (as part of his analysis of Orthodox Theory) in the following terms: "photography can be pure, or photography can be art, but it cannot be both" (Costello 2018b: 230).

In Scruton, we find the most radical form of this artistic skepticism. The ideal photograph, being purely mechanical, causal, and non-intentional, would not be art—because it would not even be a *representation*. Photography could be representational art, but at the cost of moving away from its ideal form and becoming painting "precisely through taking representation seriously" (Scruton 2008: 156). In Hopkins, we find a milder form of this skepticism, as he does not deny that photography is a representation and that it can be art. The problem with photography in its authentic form is that it would be poor as a vehicle for communicating artistically relevant thoughts.

Based on this brief analysis, we can summarize the *idées fixes* that constitute the general lines of the Orthodox Theory in relation to the answers to the four fundamental questions of the philosophy of photography:

1. Ontological question: photographs are pictures generated in a "causal," "mechanical," and "automatic" manner, untainted by the intentions and beliefs of the photographer.
2. Question of the relationship between photographs and handmade pictures: photographs and handmade pictures are two discontinuous categories.
3. Question of the epistemic status of photographic pictures: photographs afford factive experience.
4. Question of the status of photography as art: radical and moderate skepticism and non-skepticism.

[23] Perhaps a more effective way to explore the difference between these authors is to contrast the adoption of an ontology of objects (Walton) with an ontology of events or facts (Hopkins).

Additionally, as noted at various points, a fundamental trait of the Orthodox Theory is *sublimation*, which shifts the focus away from ordinary photography and leads to the idealization of pure photography. This sublimation can be found in relation to the four fundamental questions. Ontological sublimation holds that even if ordinary photographs are influenced by the photographer's intentions and beliefs, and are therefore not completely "causal," "mechanical," and "automatic," *pure* photography possesses an essential form that is free from the contamination of the photographer's intentions and beliefs. Regarding the relationship between photographs and handmade pictures, sublimation is expressed in the idea that, even if ordinary photographs result from deliberations and manipulations that bring them closer to handmade pictures (to varying degrees), *pure* photography remains discontinuous from handmade pictures. Epistemic sublimation, in turn, posits that, even if ordinary pictures are not factive because they do not meet ideal norms of accuracy, *pure* photography would necessarily be factive. Finally, aesthetic sublimation suggests that, even if ordinary photographs are considered a full form of art, *pure* photography would be incapable of being art (in Scruton's radical version) or would be a limited art form (in Hopkins's moderate version).

To overcome the four forms of sublimation to which we are led in Orthodox Theory, it will be necessary to redirect the philosophical analysis from *pure* photography to *ordinary* photography. For this purpose, we can adopt something akin to Wittgenstein's dictum: "Back to the rough ground!" (Wittgenstein 1953: §107). Returning to the rough ground means taking the "impure" photographs of ordinary practices as the object of philosophical inquiry. This also aligns us with Wittgenstein's saying that "meaning is use." The meaning of photography is not what it *naturally means* (in Grice's terms), determined by its pure and ideal ontology, but rather depends on *how it is used* in different contexts.[24]

Thus, from an ontological perspective, the problem becomes defining photography in a way that accepts it can, to varying degrees, result from the beliefs and intentions of photographers, such that photographs and handmade pictures should be thought of as continuous categories. In an epistemic sense, returning to the rough ground means understanding how the practices of using photography are structured, in a way that it can be either factive or non-factive within those different practices. Regarding art, it means understanding how photographic pictures, whether factive or non-factive, can be used in different artistic contexts.

This is not a refutation of Orthodox Theory but a change in the object of study. While Orthodoxy focuses on the ontological, epistemic, and artistic status of *pure* photography, our aim will be to understand how we use *ordinary* photography as part of our social photographic practices. And in those contexts, photography is used as a means of communication. Thus, my intention is not to argue that photography should be *defined* as a means of communication. My project is not definitional. Rather, I want to understand how we *use* ordinary photographs (which do not

[24] I employ those Wittgensteinian ideas here without any commitment that my position accurately reflects or aligns with Wittgenstein's overall philosophy.

conform to the strict standards of the Orthodox ideal of photography) within social practices, where they are frequently employed as a means of communication. The path for this shift in focus can be found in the New Theory, particularly in Dominic Lopes's version (2016, 2018), where the "social turn" plays a fundamental role.

1.5 The New Theory of Photography

The New Theory of the philosophy of photography encompasses various recent approaches that have emerged in the analytical Anglo-Saxon tradition, stemming from critiques of the central ideas of Orthodox Theory. The main contributions to the New Theory include the works of Paloma Atencia-Linares (2012, 2018b), Dominic Lopes (2016, 2018), Catharine Abell (2018), and Diarmuid Costello (2018a, b). The emergence of the New Theory can be traced back to a shift in perspective present in the work of Patrick Maynard (1997) and the developments proposed by Dawn Wilson (née Phillips 2009).

Although there are important distinctions among the proponents of the New Theory, my analyses will focus on the proposal by Dominic Lopes (2016). Lopes's theory is not free from criticism, as will be seen. The significance of Lopes's work for the purposes of this book lies in what he calls the "social turn" (2018). Lopes conceives of photography as constituted by different social photographic practices. This "social turn" by Lopes will be taken as the starting point for the social ontology of photographic practices that I will propose in the following chapters. This "social turn" will be fundamental for moving beyond the restrictions that Orthodox thought imposes on photography. Thus, it is paramount to understand how Lopes constructs his version of the New Theory and to identify the points that need revision to arrive at the conception of the "social turn" that will serve as the basis for my theory (in one important aspect, my approach will differ from Lopes's).

In general terms, concerning the ontological problem of determining the nature of photography, the New Theory operates the following shift in focus. Instead of questioning what photographic images are and what kind of relationship they have with the world, the philosophical analysis will prioritize an understanding of photography as the *process* of producing images through the use of photographic *technologies*. Thus, defining photography splits into two distinct questions: "What are photographs?" (from an Orthodox perspective) and "What are photographic technologies?" (from the New Theory perspective). This shift in focus is characterized by Atencia-Linares as follows: "one of the marks of the New Theory is to put the emphasis on the photographic process rather than on finding a definition of what a photograph is" (2018b: 217). In agreement with Atencia-Linares, Costello states: "what is most productive about New Theory is its reorientation from photographic product to process" (2018b: 233). Thus, unlike Orthodoxy, which prioritizes understanding the result of the photographic process, the New Theory seeks to understand photography as the process of producing images.

This shift in focus of the New Theory is often traced back to the following definition by Patrick Maynard. According to the author, photography is "a branching family of technologies (...) whose common stem is simply the physical marking of surfaces through the agency of light and similar radiations" (1997: 3). Photography is thus conceived as different technologies that resemble each other to varying degrees, with the common element being the physical marking of surfaces by the action of light or similar radiations. And depending on what we accept as radiations similar to light, the scope of photographic technologies would be radically expanded. This definition significantly broadens the concept of photography beyond the traditional Kodak-type paradigm embraced by Orthodox Theory. It encompasses a wide range of photographic practices, including photolithographs, X-rays, medical imaging diagnostics, solarizations, and more. In this broad perspective, notably, Maynard's definition does not mention the mimetic qualities or the causal connection between the depicted objects and the image. His definition allows for distinguishing photographic technologies that are *pictorial*, producing images of recognizable objects, and *non-pictorial* technologies, generating non-figurative differentiated surfaces by the action of light or similar radiations.

A central part of the debate among the proponents of the New Theory is determining the stages necessary and sufficient for a technology to be classified as photographic. The ontological problem of photography thus becomes the problem of defining the processes that constitute photographic technologies.

According to Dominic Lopes (2016), drawing on Dawn Wilson (née Phillips 2009), there are four stages in photographic technologies. The first stage, according to Lopes, is the *pro-photographic scene*, which is essentially any scene in relation to which some kind of photographic apparatus is set up. The second stage is the optical projection of light from the pro-photographic scene onto a photosensitive surface—referred to by Lopes as the *light image*. In traditional photography, the light image is the dynamic inverted image projected inside a camera obscura, with or without the use of a lens. The third stage is the *photographic recording event*, where the information from the light image of the scene projected onto the photosensitive surface is recorded on some storage medium. In the case of black-and-white analog photography, the third stage would be the sensitization of silver salts in the film emulsion. In digital photography, the third stage would be the recording in a data file of the binary information generated by the electrical impulses resulting from the projection of the image on the sensor. At this stage, as Lopes emphasizes, we do not yet have something worthy of being called a "photograph". We only have something that may become an image—a *latent image*. In this third stage:

> Some photons from the dynamic light image have moved some electrons in a sensor or emulsion to a higher energy state. There is no marked or differentiated surface visible to the human eye. There is no image. Unrevealed emulsion and data files on computer chips are not images. (2016: 80)

A fourth and crucial stage is required to complete the photographic process. This final stage involves using the previously recorded information as input in a process that yields a visibly differentiated surface as its output. It is at this juncture that the recorded information is transformed into a visible image. The realm of photographic technologies offers a plethora of methods for creating such visibly differentiated surfaces. Presently, this includes various types of digital displays (such as plasma, LCD, LED, etc.), inkjet printing on papers and diverse surfaces, traditional photographic processes using photosensitive papers, among others. All these manifestations are the final product (the output) of the recorded light image information from the pro-photographic scene, forming distinct branches within the family of photographic images.

We can schematically characterize the four stages that Lopes attributes to photographic technologies as:

1. The pro-photographic scene
2. The light image
3. Photographic recording event
4. The marking of the surface

Based on the understanding of these stages, Lopes proposes the following definition:

> [A] photograph is an image output by a mark-making process taking input from an electro-chemical event that records information from a light image of a pro-photographic scene. (2016: 80)

This definition will be criticized by Atencia-Linares (2018b), Abell (2018), and Costello (2018a, b). These authors argue that Lopes's definition is excessively permissive. According to Costello, the permissiveness of Lopes's position arises from the lack of constraints about "the media through which such an image need be rendered for it to count as a photograph" nor "how much information from the light image need be preserved" (Costello 2018b: 232). Thus, as long as there is a "photographic event in the causal history of an image" and "some modicum of information transfer," the resulting image would be a photograph (Lopes 2018: 228).

Costello explores the following idea to show the excessive permissiveness of Lopes's position. Suppose an artist could project a photographic image onto a canvas and outline the image with paint. Subsequently, he could use methods to blur the image (with solvent, for example) so that the result is abstract. In this case, we would have an image as the output of a process from an electro-chemical event that records information from a light image of a pro-photographic scene. But would we say that the resulting abstract painted image is a photograph? Costello and Abell oppose this conclusion. However, according to Lopes's definition, the abstract painted canvas would be a photograph because "the New Theory of photography does not require that photographic processing preserve most, or much, information recorded in the photographic event" (Lopes 2016: 123–124).

It is possible to apply different constraints to restrict the excessive permissiveness of Lopes's definition. One way, which will be criticized by Abell, would be to require that the resulting image must recognizably (or potentially recognizably) depict the pro-photographic scene involved in its production (2018: 213). But this would veto the possibility of abstract photographs. Abell suggests that the proper way to make Lopes's theory less permissive is to adopt an appreciative condition on what suffices to count as a photograph of something. Abell's appreciative condition is that the "genesis of the resultant visual image should bear on its appreciation" (2018: 213). For the photographic genesis to play a role in the appreciation of the resulting image, the resulting image must convey information from the recording event. In the previous example, the photographic genesis does not bear any role in its appreciation, as the outlining with paint and use of solvents to create an abstract image would not be effective at carrying information from the photographic recording event.

Although there are different ways to restrict the excessive permissiveness of Lopes's definition, my interest is in noting the consequences of adopting his permissive version and how the author addresses the problem of the epistemic status of photographic images that results from this permissiveness (which we will discuss next).

One interpretative gain resulting from the permissiveness of Lopes's definition is that he manages to explain how photographers have an extensive range of means to express their intentions and beliefs through the manipulation of photographic processes. The photographer can exercise their deliberations with the purpose of conveying their thoughts and intentions by manipulating the four different stages: choosing the pro-photographic scene and manipulating its elements; selecting the photographic apparatus and its application; intervening in the formation of the light image, determining its formation and properties; opting for different recording processes for varied outcomes; employing numerous methods in the process that transforms the recorded image into a visibly distinct surface, etc.[25] This allows for the possibility that photographic images can be non-factive and even fictional.[26] Photographic technologies may or may not serve as systems that preserve information from the pro-photographic scene. It all depends on how the stages of the photographic technology are operated by the image producer. Thus, an important consequence of the permissiveness of Lopes's definition is the acceptance that photographs can be non-factive and still be considered photographic images if they result from a photographic technology with the four stages.

From this perspective, we could formulate Lopes's version of the New Theory as the acceptance that photography could embody what Grice (1957) terms "natural

[25] In truth, the deliberations available to the photographer are much more complex than this description proposes. Each of these options breaks down into numerous possibilities.

[26] An important distinction that I will draw later is between Quality Contexts (in which we have Grice's Maxim of Quality in operation) and non-quality contexts. I will develop the photographic version of this distinction based on the analyses of Daisy Dixon (2018: 56–76). In Quality Contexts, it makes sense to question the accuracy and the truth or falsity of the pictorial content. Non-factive photographs in Quality Contexts would be false. Fictional non-factive images, on the other hand, would not be considered false if used in a non-Quality Context. In a non-Quality Context, there is no presupposition that the image must correspond to the world. Thus, the use of photography, in this case, does not aim to perform an assertive speech act, where the user of the image is committed to the truth of the pictorial content. Thus, fictional photographs, if used in a non-Quality Context, would not be false photographs.

meaning" and also "non-natural meaning," depending on how the technology is used. For Lopes, it would function as natural meaning when used in such a way that the resulting image provides visual information about the scene in the pro-photographic scene through *belief-independent feature tracking*.[27] In this case, the image could be used as one that affords a factive experience. However, the resulting image could still be photographic even if it does not provide visual information about the scene in the pro-photographic scene, serving as a form of *belief-dependent feature tracking*. This image would still be photographic, even if non-factive. The Orthodox Theory has traditionally resisted this dual functionality, relegating images that result from processes expressing the photographer's intention to a non-photographic realm—photography would have become "painting" (Scruton 2008). However, in Lopes's version of the New Theory, the resultant image in either scenario ought to be photographic.

To do justice to Robert Hopkins's Orthodox position, it is important to note that he would accept that photographs can be non-factive and embody what Grice terms "non-natural meaning." But if the manipulation of photographic processes concerns content-determining properties, the resulting image would no longer be *authentic*. In this respect, his position is similar to Scruton's discontinuism (2008), which understands the non-factivity of photographic images as a departure from the ideal photography. The fundamental point of difference is that, in Lopes's version of the New Theory, photography, as a technology, could be used to produce non-factive images that are still photographs, and not a separate category of images divorced from photographic images generated by epistemically reliable processes (as in the discontinuist approach).

From this perspective, the author concludes in a *continuist* tone:

> The take-home message is that photography and drawing are not disjoint. Each is not what the other excludes. They overlap in the sense that each can serve the other (Lopes 2016: 85).

Lopes's continuism is achieved through a double argument that turns Kendall Walton's discontinuist position against itself. For Walton (2008a, b), photographic images are produced by belief-independent feature tracking, while handmade images are belief-dependent. Lopes's double argument leads to the abandonment of both theses. As already seen, in his version of the New Theory, photographs can be instances of belief-dependent feature tracking. Lopes provides the following examples to demonstrate the falsity of the second thesis. He explores contexts where handmade pictures are created without the artist holding beliefs about their subject or lacking the conceptual tools necessary for such beliefs.

When Galileo first pointed his telescope at the moon, he did not have preconceived beliefs about what he would find on its surface (Lopes 2016: 76–77). His subsequent drawings of the moon, created without any prior belief about its features, led to his realization that its surface was rugged and irregular, akin to Earth's,

[27] Next, we will explore the importance Lopes places on the idea that, in epistemic contexts, photography should be produced by belief-independent feature tracking.

with mountains and craters. This understanding emerged post-drawing, indicating that the drawings were a form of *belief-independent feature tracking* of what he observed.

Another example offered by Lopes is the case of a child named Nadia (Lopes 2016: 74–77). Diagnosed with a form of autism that severely limited her motor skills and verbal communication, she demonstrated a remarkable aptitude for drawing at a young age. Her drawings, created without formal training, exhibited a level of detail and beauty likened to those of Leonardo da Vinci, as noted by a psychologist studying her (Lopes 2016: 74). Lopes argues that Nadia's initial drawings could not have been belief-dependent feature tracking, as she lacked the conceptual skills necessary to form beliefs about her subjects.

Therefore, contrary to Walton, according to Lopes, both in the case of Galileo and Nadia, the drawings could not be belief-dependent feature tracking. Thus, drawings and photographs are not fundamentally different in this respect; both can be either belief-independent or belief-dependent.

However, the permissiveness of Lopes's approach leads to the following problem. If photographs can be belief-dependent feature tracking and thus fictionally competent, how do we explain our commonsense intuition that photographs hold epistemic superiority over handmade pictures? Lopes posits that the belief in the epistemic superiority of photography is a *psychological fact*, one that Orthodox Theory tries to elevate to a *normative claim* defining photography (Lopes 2016: 106). Lopes will accept this psychological fact and reformulate the normative claim. The way Lopes resolves this problem will characterize the "social turn" in his theory.

As seen in the previous section, Robert Hopkins is one of the authors who most vehemently argues for the normative claim that it is inherent to photography to have an epistemic virtue not possessed by handmade pictures (Hopkins 2010). According to him, when photography is operating properly, it affords factive experience, since providing accurate pictorial experience is an "overarching goal of the design" (Hopkins 2010: 9). Photography is only non-factive when the mechanism fails to function as designed, whether due to technical malfunction or human interference. Thus, for photography, factivity would be a matter of necessity, as the photographic system itself pursues the goal of providing factive experience when operating properly.

Lopes's central idea is that, contrary to Hopkins's Orthodox perspective, the epistemic value of an image as a vehicle for knowledge about the world is neither intrinsic to photographs nor extrinsic to handmade drawings. Drawings can be reliable vehicles of information about the world, and photographs unreliable ones. What guarantees reliability is not the type of image but the *practices* and *rules* that guide different uses in different *social contexts*. What makes the processes of producing images reliable (whether photographic or handmade) are the *social practices* that govern their production and use. It is this importance placed on the idea of social practice that constitutes the "social turn" in his theory. (This idea will have profound significance for the position that will later be proposed in this book). As Abell notes, in Lopes's approach, "belief-independence isn't built into the nature of photography but imposed on it from the outside by social norms" (Abell 2018: 215).

This is the juncture where, in my opinion, Lopes makes his most outstanding contribution to the ongoing debate about photography. In Lopes, the branching element outlined in Maynard's definition has a social dimension. Photographs can be factive or non-factive depending on how the four constituent stages are employed in accordance with the goals of the social practice. In epistemic contexts, strict rules govern the photography production to ensure its role as a reliable source of information. Conversely, in non-epistemic practices, such stringent regulations may not apply. In his book from 2016, Lopes does not give us a theory of how those social practices are structured and how the rules determine the status of the photographs that result from the use of photographic processes within those practices. (I will do that later in this book). He only illustrates this with examples across different fields, highlighting how production rules strengthen the epistemic reliability of photographs.

For Lopes, when images depict by belief-independent feature tracking, they are more likely to be reliable sources of knowledge about the world. In scenarios where an image is based on beliefs or is fictional, it is less likely to be a reliable source. These two paths would be open to both photographs and handmade images. Lopes's continuist position can be summarized in the terms proposed by Costello:

> Belief-independent feature tracking does not parse between photography and painting, but between the automatic and non-automatic processes to be found in both. (Costello 2018a, b: 87)

Adherence to technical standards is essential for various photographic equipment and processes to produce images that aim at giving visual information about the objects they depict. The development and refinement of photographic equipment and methods over more than a century have been frequently driven by their teleological function of conveying knowledge. Using this equipment in line with technical standards typically results in images with mimetic qualities relative to the objects photographed, thereby facilitating the recognition of the objects' visual properties. In such instances, subject to the sensitivity limitations of the equipment, there exists a counterfactual dependence between the photographed scene and the resulting image. Yet, *this is not the essence of photography*; it represents merely one set of practices among many, shaped by specific techniques and tools aimed at knowledge acquisition.

In fields like medicine, science, and forensic photography, explicit rules dictate the type of equipment and techniques used, aimed at guaranteeing that the resulting images are reliable sources of knowledge. For instance, in forensic photography, guidelines set by the United States National Institute of Justice specify the photographic methods and equipment, with non-compliance potentially disqualifying a photograph's use in legal contexts (Lopes 2016: 109–110). In photojournalism, organizations like the Associated Press enforce strict policies against altering photographic content, with violations potentially leading to severe professional repercussions (Costello 2018a, b: 152). Ignoring this rule by a photojournalist can lead to job termination and ostracism, as evidenced by several cases. It is the adherence to the rules determined by the contextualized practice of photography that allows us to use it as a source of knowledge about the world. Thus, according to Costello, in Lopes's "social turn," "certain institutional contexts (such as forensic, legal and press) work to ensure the preservation of belief-independent feature tracking, though this is neither unique nor intrinsic to photography" (Costello 2018b: 232).

In sum: photography's different social functions, techniques, and equipment yield diverse outcomes, each with varying degrees of epistemic reliability. Some photographic social practices may be categorized as using photography as belief-independent feature tracking, while others may not. Lopes's version of the New Theory posits that all these varied uses and practices, despite their differences, can be equally considered photographic. As summarized by Lopes, his aim "is to break the dominance of the epistemic conception of photography over how we think about its artistic promise" (Lopes 2016: 113).

But Lopes's idea extends beyond photography. He points out that similar rule-based practices exist in contexts where drawings are used, such as in architectural plans, civil construction projects, topographic maps, and scientific illustrations. These drawings, governed by specific rules, can also serve as reliable sources of information about the world. For instance, in civil construction, adherence to image production rules is critical, with deviations potentially leading to catastrophic outcomes.

We can then conclude that the main difference between Hopkins and Lopes is that, for the former, the norms that guarantee the epistemic superiority of photography are part of the photographic system, while for Lopes, these norms are determined by the social practice of use. Thus, in Hopkins's view, photography itself carries the "background goal of supporting accurate seeing-in" (Hopkins 2010: 12). Although Lopes recognizes that the equipment is designed with a certain purpose in mind, the epistemic purpose is not internal to the photographic system but imposed on photography by certain social practices. There are practices, as Lopes exemplifies through Lomography, whose equipment is not designed with factivity and/or accuracy in mind. This allows us to note that, according to Lopes, factivity is not intrinsic to photography—although it is a common goal shared by many social photographic practices.

Thus, the difference between Hopkins and Lopes's version, which I will explore in the following chapters, is that for Hopkins, it is possible to speak of a *purpose* of photography independent of social practices of use. The intention of photography to provide factive experience is, in Hopkins's view, part of photography's self-conception. Thus, the *correct* use of photographic equipment (when it is working properly) guarantees factivity. The position I will adopt, following Lopes, is that purpose only exists within social practices, as purpose depends on the collective intentionality of the participants in the practice. Only in relation to the internal purpose of a practice could we speak of correct or incorrect use. We can speak of the purpose of the photographic system only in a derivative sense, based on the intentions or purposes of the participants within a particular photographic practice. Moreover, the same equipment can be used in different social practices with different purposes.

In non-epistemic artistic practices, the use of the equipment could only be described as "wrong" if we presuppose as "correct" the use determined in certain epistemic practices. But, by accepting a certain incommensurability of social practices (which I take to be present in Lopes), the use of equipment in non-epistemic artistic practices would be *correct* in relation to the practice. Things are working properly, even if the resulting image is non-factive, in an artistic non-epistemic photographic social practice. Moreover, as will become evident in later chapters, different epistemic practices can have varying levels of accuracy, such that an image deemed "factive" in a permissive context may be accepted as "non-factive" in a more restrictive epistemic context. I will argue that "factivity," "accuracy," and "truth" only have meaning within an epistemic practice (which I will refer to as assertive, using John Searle's taxonomy in Sect. 3.5.1 and Chap. 4), which determines what type of visual information from a photographic image the practice considers relevant, what degree of accuracy the practice requires in correspondence with the world, and what method it uses to determine this correspondence.

The argument put forward by Orthodox authors, which takes photography to be necessarily factive, often draws a parallel between the supposed factivity of memory and the factivity of photography. Hopkins resorts to this parallel by stating that, similar to photography: "If you remember that your mother was born in 1940, then your mother was born in 1940. Again, you might be mistaken, about this as about any other matter. But, if you are, that just shows that your state of mind does not count as remembering" (Hopkins 2010: 3). From this perspective, cases of misremembering would not be strictly memories. However, as cases of misremembering are common, several authors specializing in the debate on episodic memory adopt the view that memory can signify different things in different contexts and accept that, in many cases, it is not factive (McCarroll et al. 2022; Openshaw 2023).

Simulationism considers that memory and imagination belong to the same natural kind (Michaelian 2016), and incorrect representations of the past should be considered as memory (de Brigard, 2014). Even causal theories of memory nowadays are tending to reject factivity (Robins 2020). Few defend the factivity of memory, such as Bernecker (2017).[28] But even if we accepted that episodic memories are factive, the parallel with photography conceals an important and crucial difference. It is possible to argue that memory is a *natural kind*, so if memory has a certain purpose, this would be internal to the nature of memory. However, photography is not a natural kind but a *technology* developed by humans for certain ends. The purpose of photography is not part of its nature but determined by the intentions of those who produce and use it.

In summary, the Orthodox position not only reifies Kodak-type photography but also epistemic practices as something that would determine the essence of photography. This can be explained by the historical contingencies that determined the

[28] I am immensely grateful to my colleague José Carlos Camillo Castro Neto (Ph.D., UFG) for clarifying these points regarding the most recent bibliography on episodic memory and for providing me with a (forthcoming) paper of his authorship that maps the current debate on simulationism.

social photographic practices we prioritize. However, it is not difficult to imagine another set of historical contingencies, such that we would live in a world where photography is used exclusively for non-epistemic artistic purposes. In this world, equipment could be developed to facilitate the introduction of elements and manipulate the resulting image in such a way as to not provide accurate information about the world, aiming for purely aesthetic and artistic goals. From this perspective, the less factual the resulting image, the better. The equipment would be malfunctioning when it provided merely factual images. It could be argued that such a world would involve a *self-conception* of photography quite different from our current understanding. But I believe this is the world we are soon approaching, with the advancement of digital photography and the incorporation of image editing tools via Artificial Intelligence in the photographic equipment itself. We increasingly value equipment that allows us to alter the determining properties of pictorial content more simply and effectively. Just as modernist painting abandoned the pretense of being a mere copy of the world, this is the rupture we are currently witnessing concerning photography. Especially when we take self-portraits (the (in)famous *selfies*), we do not want the photograph to portray the world as it is, but as we would like it to be. Nowadays, in various contexts where photographs are used, we have reached the point where the relationship between pictures and the world has been reversed. With the widespread availability and constant use of digital image manipulation techniques, many young people now seek out doctors from various specialties (such as dermatologists and surgeons) to alter their faces and bodies to match how they appear in photographs. In these cases, the world is reshaped to faithfully mirror the photographs. However, while these manipulation techniques are now easily accessible to everyone, they were already available to specialists in analog processes, albeit requiring much more effort.

But there is an element of Lopes's position that does not seem correct to me, or at least requires some caution. Lopes will reformulate the normative claim—encompassing both photographic and handmade images—in the following terms:

> The improved normative claim is that images made by belief-independent feature-tracking have an epistemic virtue not possessed by other images. Epistemically virtuous images come not from photography but from belief-independent feature-tracking (2016: 112).

The point observed by Abell and Costello, with which I agree, is that "belief-independent feature tracking should occupy a 'more modest place in an explanation of photography's epistemic value'" (Abell 2018: 215). As Costello notes, "flouting belief-independent feature tracking may sometimes aid our epistemic enquiries, by making it easier to isolate and extract the sought after information" (Costello 2018a, b: 232). This would be common, according to Costello, in medical imaging and astrophotography. The example of astrophotography is interesting because Lopes himself explores it and argues very closely to the considerations made by Costello—which will force us to rethink Abell and Costello's criticism and the very formulation of Lopes's improved normative claim. Many astrophotographs are made using radiation in the non-visible spectrum. The image is mapped into the visible spectrum (receiving a *false color*), often guided by the astronomer's aesthetic interests. There is then a belief-dependent dimension in the process of producing these images that use false color. But, as Lopes acknowledges, “false color images are perfectly good sources of information” (2018a, b: 131). Certainly, they would not be good sources of information about the colors of constellations and stars, but they remain good sources of many other kinds of information. Another example mentioned by Abell and Costello is medical imaging, in which doctors manipulate color and other values to facilitate diagnosis since it “aids extraction of the sought-after information” (Costello 2018a: 135). These would be distorted pictures (in a sense “inaccurate”), but with this distortion, they can more immediately and accurately fulfill their epistemic function.

We could be charitable to Lopes and say that he partially accepts Abell and Costello’s criticism. Photographs can be good sources of information, even if they are belief-dependent. But, considering Lopes’s improved normative claim, we would have to conclude that they are perfectly good sources of information *despite* being belief-dependent. Furthermore, for Lopes, what would guarantee that they are good sources of information are the aspects of the image produced by belief-independent feature tracking.

In the following chapters, I will adopt and develop Lopes’s “social turn” to propose an ontology of social photographic practices, with the aim of understanding how the collective intentionality of participants determines the purpose of photographs within a practice, and how this purpose can be related to constitutive and regulative rules. This will allow us to understand what must be shared by users and viewers for communication via photographs to occur within practices we can describe as conventional contexts.

But I will diverge from Lopes regarding the relationship between the epistemic value of photography and its production as belief-independent feature tracking. Photographs classified by Lopes as belief-dependent feature tracking could nevertheless be perfectly good sources of information. As mentioned earlier, I support the idea that “factivity,” “accuracy,” and “truth” only have meaning within epistemic practices, which determine what kind of visual information from a photographic image is relevant to the practice, the degree of accuracy required in correspondence with the world, and the method to be used to determine this correspondence. This includes the possibility that in certain practices we accept as epistemically reliable photographs that are belief-dependent feature tracking.

1.6 Beyond the Orthodox Theory of Photographic Communication

We can now envisage a broader approach to thinking about photographic communication, taking Robert Hopkins as a starting point and incorporating the critiques of Paul Grice’s theory along with the developments proposed by the New Theory. This outline of a broader communicative theory of photography will enable us to identify the problems that must be addressed in detail for the foundation of a general theory of ordinary photographic communication (in the following chapters).

My intention is not to refute Hopkins but to operate a shift in perspective and scope. By changing the perspective and scope, we could then incorporate his various contributions into a more general theory. This shift can be carried out by rejecting four constraints that characterize his theory:

1. The restriction to authentic photography.
2. The restriction to the artistic context.
3. The acceptance of the restrictions imposed by Paul Grice on communication.
4. The restriction of communication to the producer-spectator perspective.

These four constraints will be overcome through four expansions:

1. Acceptance of the general outlines of the New Theory.
2. Analysis of ordinary contexts of use (whether artistic or not).
3. Expansion of Grice's theory in light of the critiques by Recanati, Sperber and Wilson, and Moore, along with the inclusion of *speech act theory*.
4. Expansion of communication to the user-spectator perspective.

Let us examine in greater detail how these expansions will be addressed and some of their consequences. This will allow us to summarize the ideas presented in the previous chapters and determine the fundamental difficulties that must be overcome in constructing a general theory of photographic communication.

The centrality that Hopkins attributes to authentic photography clearly situates him within the scope of the Orthodox Theory. His ontology of authentic photographs is based on the idea that photographs are pictures produced via a chain of mind/belief-independent sufficient causes. Consequently, he assumes a discontinuous stance in relation to handmade images, which are considered mind/belief-dependent. From an epistemic perspective, this discontinuity is founded on the notion that photographs afford factive experience, whereas handmade images are non-factive. Factivity is seen as a necessary trait of authentic photography and results from photography's inherent goal of providing accurate depictions of the world.

In the New Theory, on the other hand, photography is conceived as a family of technologies for surface marking by the action of light or similar radiations (Maynard 1997). In Lopes's version (2016), photographs can be belief-independent or belief-dependent, depending on how the technology is used and regulated within social practices. Thus, photographs and handmade images are best understood as existing along a *continuum*. The factivity of photography would be restricted to its use within epistemic social practices, where the production of the image is regulated. Non-epistemic practices would be equally photographic, even if the resulting image is not factive (fictional or false).

Lopes's version of the New Theory seeks to break the dominance of the epistemic conception of photography and offers an excellent framework for considering ordinary contexts of photographic use (whether artistic or not). Ordinary practices can be divided into epistemic and non-epistemic, as well as artistic and non-artistic.[29] There could be practices that are both epistemic and artistic (such as *highbrow* photojournalism—for example, Cartier-Bresson and Sebastião Salgado) and practices that are simultaneously non-epistemic and non-artistic. And there would be a *continuum* between all four possible combinations.

As discussed in the previous chapter, Hopkins (2015) attributes to photography the background goal of supporting accurate seeing-in. As a result of this goal, for Hopkins, when the photographic systems are functioning *correctly*, the resulting images afford a factive experience *out of necessity*. For Dominic Lopes (2016, 2018), on the other hand, factivity is not intrinsic but is imposed on photography by certain social practices. I wholeheartedly agree with Lopes on this point. The attribution of intention or purpose to photographic systems depends on the intentionality and purpose of the system users. This intentionality, in turn, can be considered in relation to different social practices and, therefore, is dependent on the *collective intentionality* of the participants in the practice (we will explore this idea in Chap. 4). We can only speak derivatively of the intention or purpose of the photographic system based on the intention or purpose of the participants in a particular photographic practice.

The third point of expansion concerns the way Hopkins bases his theory on Grice's theory of communication. For Grice, the intentional use of *natural meaning* would not constitute genuine communication because what is meant could be directly inferred from what the expression naturally means. The interlocutor would not necessarily grasp the communicated thought due to the recognition of the *communicative intention*. And the same would apply to photographs, as paradigmatic examples of natural meaning. The acceptance of these ideas leads Hopkins to the conclusion that, for there to be communication via photographs, the elements the photographer uses in producing the image "must lie under the artist's control in a way that appreciators can detect" (2015: 336). But what an authentic photograph naturally means would be independent of the artist's control. More precisely, the putative fact depicted in the photograph is independent of the photographer's communicative intention. Photographic communication would be restricted to vehicle properties that the photographer could use as *non-natural* meaning, whose modifications do not preclude it from being an authentic photograph.

We have seen that several authors oppose the thesis that communication is never based on natural meaning, notably Recanati (1986) and Sperber and Wilson (1995). The acceptance of certain central elements in these critiques allows us to broaden the scope of the theory of photographic communication. These authors accept the existence of two distinct cases: there are cases in which what is communicated depends solely on the speaker's intention (as in the use of non-natural meaning), and cases in which what is uttered provides additional evidence to what is intended by the speaker (as in the use of natural meanings). And there is a continuum between

[29] Drawing from John Searle's taxonomy of speech acts (1979), in Sect. 3.5, we will be able to distinguish practices in relation to their *illocutionary purposes*, moving beyond this initial characterization developed based on Lopes.

these two extremes. But, regardless of whether what is uttered provides additional evidence, communication would have two levels. As Sperber and Wilson (1995) describe, communication comprises a foundational layer of information about any subject (the *informative intention*) and a second layer indicating that the first layer was intentionally manifested (the *communicative intention*).

We can then conceive of the communicative use of photographs as encompassing, in addition to the modes described by Hopkins, the possibility of being constituted by the ostensive-intentional use of photographs as a foundational layer of information. The *communicative intention* would be the intentional use of the photograph to communicate something to someone through the recognition by the spectator that the photograph was intentionally used. This includes the manifestly intentional uses of both authentic and inauthentic photographs (thus falling into what Grice calls natural or non-natural meaning).

The fourth point of expansion that I will adopt concerning Hopkins is the abandonment of the centrality of the producer-spectator perspective, which becomes possible as a consequence of the first three developments. Hopkins sought to understand how the artist, as a producer of photographic images in an artistic context, could use authentic photographs to communicate artistically relevant thoughts. By abandoning the restrictions of authentic photography, the artistic context, and the restricted reading of Grice's theory, we can think of photographic communication as primarily based on the ostensive-intentional act of proposing a photograph as a candidate for attention in an ordinary (not necessarily artistic) interlocutory context, whether the user of the photograph is its producer or not. The producer-spectator communication would be a particular case of the general theory of user-spectator communication.

This change of perspective will have an important consequence (previously mentioned in Sect. 1.2). From a temporal perspective, Hopkins's approach to photographic communication requires that the producer have the intention to communicate a particular thought *at the moment* of image production (or *at an earlier moment*), and this intention must be made manifest through the interplay between the vehicle properties and the content. Thus, the communicative intention would be determined by the original context of image production and the producer's intentions. As previously mentioned, I believe that this approach risks leading us to a *mythology of the artistic creation process*, which presupposes that the artist knows what they want to communicate *at the moment* of image production. Adopting the user-spectator perspective shifts the temporal focal point of communication from the act of *production* to the moment of *use*. This allows us to abandon the notion that the temporal context determining what is communicated is solely the original context of production. We can also abandon the idea that the intention determining what is communicated is the communicative intention of the *producer*. For communication to occur, it is necessary that, in the act of *using* the image, the *user* makes their communicative intention manifest and that it is apprehended by the spectator.[30] This approach offers greater flexibility in thinking about photographic communication, as the same image could be used with different communicative intentions in different contexts by different users, resulting in the communication of different thoughts, whether or not these thoughts align with the communicative intention of the image's producer at the time of its production.

We can then expand the modes of photographic communication, adding these expansions to Hopkins's ideas. From Hopkins's perspective, communication centers on the use of vehicle properties in relation to a certain pictorial content, with the intention of conveying a thought. Among the properties relevant to picture appreciation, the author highlights *content-determining* properties and *local* and *global content-neutral* properties. According to him, photographic communication also relies on *what the image depicts*, its *visual properties*, *how it depicts* them (including *geometry*), as well as the *means of production*. I will call the elements that are directly or indirectly related to the pictorial content *intra-pictorial elements of photographic communication*. The ones *directly* related are content-determining properties, visual properties, and geometry, while the *indirect* pictorial elements are local and global content-neutral properties and means of production. The intra-pictorial elements that are directly related to pictorial content can be broadly described as *design properties*. Thus, we can summarize Hopkins's ideas in the following way:

> *Intra-pictorial elements of photographic communication*: design properties, content-neutral properties, and means of production.

The design properties are the ones directly related to determining the first layer of photographic communication:

A. *What the image depicts.*

However, as seen in the analysis of the example used by Hopkins of Shomei Tomatsu's photograph "Beer Bottle after the Atomic Bombing" (Museum of Modern Art, New York), there is a crucial difference between *what the image depicts* (in the case of Tomatsu's work, a twisted and charred bottle) and *what the use of the picture communicates* (something about "the effects of the bombing on its victims"). Thus, in addition to what the image depicts, there is:

B. *What the use of the picture communicates.*

In relation to Sperber and Wilson's (1995) perspective on Grice's theory of communication, what the photographic image depicts (determined by its intra-pictorial elements) can be considered the foundational layer of information (the *informative*

[30] We will see in Chap. 4 that, in conventional contexts, this centrality of reflexive intention can be sidestepped or even completely disregarded, as intending the illocutionary purpose of a photographic social practice will be a necessary condition for participation. As a result, this intention can be attributed to all images used within the context, regardless of whether the user of the image makes this intention ostensively manifest.

intention), upon which the *communicative intention* operates as an ostensive-intentional act. The central problem of the theory of photographic communication will be precisely to explain how we move from (A) to (B)—from (A) what the image depicts to (B) what its use by someone in a given context communicates.

Moreover, we have seen that the same photographic image, if it were the work of a different photographer in a different context or use, would communicate something different, even if it had the same design, content-neutral properties, and means of production.[31] The idea that photographic communication is reduced to the manipulation of intra-pictorial elements is a reduction of communication to *pictorial communication*. But this is only the first layer of photographic communication. The same image, with the same intra-pictorial elements, can be used for different communicational purposes, by different people, in different contexts. This allows us to notice that the transition from what the image depicts to what it communicates does not depend solely on *intra-pictorial* elements but also on *extra-pictorial* elements. Extra-pictorial elements are as important to photographic communication as the intra-pictorial elements (if not more so).

Extra-pictorial elements of communication: who uses the image, when, and in what context.

We then have the following structure of communication, in which we articulate Hopkins's ideas and the aforementioned additions:

A. What the image depicts—determined by intra-pictorial elements.
B. What the use of the picture communicates—determined by extra-pictorial elements.

Moreover, as we will explore in detail in Chap. 2, the transition from intra-pictorial elements (such as design and vehicle properties) to pictorial content will depend on various context-sensitive factors. Thus, what a picture is used to *depict* (including pictorial content, denotation, and expressive properties) will depend on context-sensitive components. This allows us to recognize that for photographic pictures to function as a means of communication, both the user and the viewer must share the context-sensitive determinations that define what the picture is used to depict and denote within that specific context of use. In technical terms, we will conclude by the end of Chap. 2 that the same intra-pictorial elements can be employed to convey different *propositional contents*, and the user and the viewer must be in sufficient agreement about this propositional content for the photograph to serve as the foundational layer of information.

But there are a series of questions that must be answered for these ideas to be developed into a general theory of photographic communication.

[31] In this case, I am considering the pictorial content as the scene depicted by the image, without considering the denotation. Denotation, as we will see in Sects. 2.2 and 2.3, is not part of the pictorial content but results from the different modes of mapping the pictorial content onto the domain of existing objects. We could consider what the picture *depicts* as encompassing pictorial content and denotation. But, this, in turn, will depend on the speech act intended by the image user or the constitutive rules of the context of use (in the case of conventional contexts).

- How do design and vehicle properties determine the pictorial content?
- How does the pictorial content relate to what the use of the image communicates?
- How does communicative intention relate to and determine what the picture communicates?
- How do variations concerning users and/or contexts impact the determination of what is communicated?

The way I will address these questions in the following chapters is by exploring the parallel between the contextualized use of photographs and *speech act theory* (originating from John Austin (1962) and John Searle (1969, 1979)). The parallel between picture and speech act is not a novelty in the studies of pictures and communication. The novelty of my contribution to the debate lies in the development of these ideas in relation to photography and in the proposal of an ontology of social photographic practices, along with a taxonomy of photographic social practices. This framework deepens the current understanding of the philosophy of photography as presented in Lopes (2016).

In speech act theory, there is a fundamental distinction between what is said through an utterance (the *locutionary act*) and the action performed through what is said (the *illocutionary act*). I propose that, in communication via photographs, what the image depicts (A) functions as the *locutionary act*, while what the use of the picture communicates (B) corresponds to the *illocutionary act* performed through the contextualized use of the picture. Certainly, this parallel is not without problems (as we will see in Sect. 2.5). We can also understand how the vehicle properties are related to the *utterance act*, which determines the photographic *locutionary act*. In turn, we can also analyze the effects of the image's use in relation to its *perlocutionary effects*.[32]

The connection between speech act theory and the developments from Grice's theory of communication can be established through the parallel between Sperber and Wilson's two fundamental layers: the informative intention and the communicative intention. These correspond to the *locutionary* and *illocutionary* acts. From this viewpoint, we can identify the two fundamental layers of photographic communication:

A. Informative intention—Photographic Locutionary act
B. Communicative intention—Photographic Illocutionary act

The development of this central idea will provide us with the following structure of photographic communication, which is isomorphic to speech act theory:

[32] In Sect. 3.2.1, we will address in detail the problem of applying the expression "locutionary act" to the case of pictures. As Novitz argues, John Austin (1962) would accept the possibility of speech acts performed without the use of linguistic expressions—we can, for example, place a bid at an auction simply by raising a hand. This, in turn, opens the way, as Novitz suggests, for the possibility of performing speech acts through the use of pictures.

1. Photographic Utterance Act: The user presents intentionally a specific photograph (with particular intra-pictorial properties) as a candidate for attention to a spectator in a given context.
2. Photographic Locutionary Act: (In performing the utterance act), the user depicts a scene in which objects are spatially arranged and endowed with visual and expressive properties—(A) what the image depicts.
3. Photographic Illocutionary Act: (In performing the locutionary act) the use of the image carries out an action in that context—(B) what the use of the picture communicates.
4. Photographic Perlocutionary Act: (In performing the illocutionary act) the user affects the spectator in a particular way.

A significant portion of the analyses in the following chapters will aim to explain how these layers relate to each other and are connected to the context of use within different photographic social practices.

References

Abell, C. 2018. Out with the old? The new theory of photography. *Aisthesis: Pratiche, linguaggi e saperi dell'estetico* 11 (2): 207–216.

Atencia-Linares, P. 2012. Fiction, non-fiction, and deceptive photographic representation. In *The media of photography. Special issue of Journal of Aesthetics and Art Criticism*, ed. D. Clo and D. M. Lopes, vol. 70. Winter 2012.

Atencia-Linares, P. 2018a. Is photography really limited in its capacity to communicate thought? A response to Hopkins. *Laocoonte: Revista de Estética y Teoría de las Artes*83–96.

Atencia-Linares, P. 2018b. Why grey is the new black. *Aisthesis. Pratiche, linguaggi e saperi dell'estetico* 11 (2): 216–226.

Austin, J. L. 1962. *How to do things with words*. London: Oxford University Press.

Bach, K., and R. Harnish. 1979. *Linguistic Communication and Speech Acts*. Cambridge, MA: MIT Press.

Bernecker, S. 2017. Memory and truth. In *The Routledge handbook of philosophy of memory*, ed. S. Bernecker and K. Michaelian. Routledge, Taylor & Francis Group.

Borg, E., A. Scarafone, and M. Shardimgaliev. 2021. *Meaning and communication*. Internet Encyclopedia of Philosophy. Available at: https://iep.utm.edu/meaning-and-communication/.

Costello, D. 2018a. *On photography: A philosophical inquiry*. New York: Routledge.

Costello, D. 2018b. Whither photography theory: Replies to Abell and Atencia-Linares. *Aisthesis: Pratiche, linguaggi e saperi dell'estetico* 11 (2): 230–234.

Danto, A. C. 1981. *The transfiguration of the commonplace: A philosophy of art*. Cambridge, MA: Harvard University Press.

De Brigard, F. 2014. Is memory for remembering? Recollection as a form of episodic hypothetical thinking. *Synthese* 191 (2): 155–185.

Dixon, D. 2018. Alterpieces: Artworks as shifting speech acts. PhD Dissertation, University of Cambridge.

Dretske, F. 1984. Abstract of comments: Seeing through pictures. *Noûs* 18 (1): 73–74.

Faria, P. 2010. Memory as acquaintance with the past. *Kriterion, Belo Horizonte* 121:149–172.

Fodor, J. 2012. Déjà vu: All over again: How Danto's aesthetics recapitulates the philosophy of mind. In *Danto and his critics*, ed. Rollins, 2nd ed. Oxford: Blackwell Publishing.

Greenberg, C. 1959. *Modernist painting*. Voice of America.

Grice, P. 1957. Meaning. *The Philosophical Review* 66 (3): 377–388.
Grice, P. 1989. *Studies in the way of words*. Cambridge: Harvard University Press.
Hopkins, R. 2005. Aesthetics, experience, and discrimination. *Journal of Aesthetics and Art Criticism* 63 (2): 119–133.
Hopkins, R. 2010. Factive pictorial experience: What's special about photographs? *Noûs*1–23.
Hopkins, R. 2015. The real challenge to photography (as Communicative Representational Art). *Journal of the American Philosophical Association* 1 (2): 329–348.
Kulvicki, J. 2006. *On images: Their structure and content*. Oxford: Clarendon Press.
Lopes, D. 2004. *Understanding pictures*. Oxford: Oxford University Press.
Lopes, D. 2016. *Four arts of photography*. Oxford: Wiley-Blackwell.
Lopes, D. 2018. Go social: Replies to Catharine Abell and Paloma Atencia-Linares. *Aisthesis: Pratiche, linguaggi e saperi dell'estetico* 11 (2): 227–229.
Maynard, P. 1997. *The engine of visualization: Thinking through photography*. Cornell University Press.
McCarroll, C. J., K. Michaelian, and B. Nanay. 2022. Explanatory Contextualism about episodic memory: Towards A diagnosis of the Causalist-Simulationist debate. *Erkenntnis*.
Michaelian, K. 2016. *Mental time travel: Episodic memory and our knowledge of the personal past*. Cambridge: MIT Press.
Michaelian, K., D. Perrin, and A. Sant'Anna. 2020. Continuities and discontinuities between imagination and memory: The view from philosophy. In *The Cambridge handbook of the imagination*, ed. A. Abraham, 293–310. Cambridge: Cambridge University Press.
Moore, R. 2017. Convergent minds: Ostension, inference and Grice's third clause. *Interface Focus* 7 (3):
Neale, S. 1992. Paul Grice and the philosophy of language. *Linguistics and Philosophy* 15:509–559.
Novitz, D. 1977. *Pictures and their use in communication: A philosophical essay*. Springer.
Openshaw, J. 2023. (In defence of) preservationism and the previous awareness condition: What is a theory of remembering, anyway? *Philosophical Perspectives* 37 (1): 290–307.
Perrin, D. 2012. Qu'est-ce que se souvenir? Vrin.
Phillips, D. 2009. Photography and causation: Responding to Scruton's Scepticism. *British Journal of Aesthetics* 49 (4): 327–340.
Recanati, F. 1986. On defining communicative intentions. *Mind and Language* 1 (3): Autumn.
Robins, S. 2020. Mnemonic confabulation. *Topoi* 39 (1): 121–132.
Russell, B. 1984. *Theory of knowledge: The 1913 manuscript*. London: George Allen & Unwin.
Schiffer, S. 1972. *Meaning*. Oxford: Oxford University Press.
Scruton, R. 2008. Photography and representation. In *Photography and philosophy: Essays on the pencil of nature*, ed. S. Walden, 138–166. Oxford: Blackwell Publishing.
Searle, J. 1969. *Speech acts: An essay in the philosophy of language*. Cambridge: Cambridge University Press.
Searle, J. 1979. *Expression and meaning: Studies in the theory of speech acts*. Cambridge: Cambridge University Press.
Sperber, D., and D. Wilson. 1995. *Relevance communication and cognition*. Oxford: Blackwell Publishers Ltd.
Sperber, D., and D. Wilson. 2002. Pragmatics, modularity and mind-reading. *Mind & Language* 17.
Sperber, D., and D. Wilson. 2004. Relevance theory. In *The handbook of pragmatics*, ed. L. R. Horn and G. Ward, 607–632. Oxford: Blackwell.
Strawson, P. F. 1964. Intention and convention in speech acts. *The Philosophical Review* 73 (4): 439–460.
Walton, K. 2008a. Transparent pictures: On the nature of photographic realism. In *Photography and philosophy: Essays on the pencil of nature*, ed. S. Walden, 14–49. Oxford: Blackwell Publishing.
Walton, K. 2008b. Meditations on a hobby horse. In *Marvelous images: On values and the arts*. Oxford: Oxford University Press.
Wittgenstein, L. 1953. Philosophical investigations. Trans. G.E.M. Anscombe. Oxford: Blackwell.

Chapter 2
Near-Side Pictorial Pragmatics Applied to Photography

Abstract This chapter introduces a key distinction between near-side and far-side pictorial pragmatics. Near-side pictorial pragmatics concerns how contextual factors determine what a picture depicts and denotes, whereas far-side pictorial pragmatics examines the speech acts performed through the use of the picture. The objective of this chapter is to apply the concept of near-side pictorial pragmatics to photography. There are various context-sensitive elements that influence how the design of a picture is mapped onto a domain of meanings, thereby determining its pictorial content. Similarly, there are also contextual factors that determine how pictorial content is mapped onto the domain of existing objects. We will examine how these mechanisms operate in photographic contexts, in order to clarify several widespread misconceptions about photographic denotation. Additionally, we will also analyze context-sensitive elements that determine the mental states a picture may express. We will also examine how John Kulvicki extends David Kaplan's indexical theory to pictures, allowing us to observe how the relationship between pictorial content and propositional content is shaped by context. I will argue that both the user and the viewer of the picture must share these elements for a photograph to be used as a foundational layer of information in a communicational context.

Keywords Photographic communication · Pictorial content · Context sensitivity · Propositional content

In the previous chapter, we extended Robert Hopkins's (2015) perspective on photographic communication through the following additions: incorporating some critiques of Paul Grice's theory of communication, abandoning the restriction to artistic contexts, expanding communication to include the user-spectator perspective, and adopting the philosophical standpoint of the New Theory. These additions allowed us to recognize two fundamental layers of photographic communication, each with determining elements:

G. Ghisoni da Silva, *The Philosophical Foundations of Photography as a Means of Communication*, Synthese Library 526,
https://doi.org/10.1007/978-3-032-02649-1_2

A. What the image depicts.
B. What the use of the picture communicates.[1]

For the subsequent developments in this book, I will explore the parallel between these layers and *speech act theory*: what the image depicts (A) can be understood as playing the logical role of the *locutionary act*, while what the image communicates (B) corresponds to the *illocutionary act* performed through the contextualized use of the picture. I will argue, in Chaps. 3 and 4, that the transition from (A) to (B) depends primarily on the rules that structure each photographic social practice—varying with different levels of conventionality and regulation. It is primarily within social photographic practices that we use photographs as a means of communication. Thus, photographic communication should be primarily understood in relation to the function, purpose, and rules of social practices. These ideas will be presented as a development of the "social turn" proposed by Dominic Lopes (2016) in his version of the New Theory.

But before reaching this level of generality, it is necessary to understand the first layer of photographic communication. The problem addressed in Chap. 2 is understanding how content-determining properties specify what the picture depicts and to what extent this is context-sensitive. These understandings require us to explore some traditional fields of picture studies that apply to images in general, not necessarily photographic ones. These studies include *picture theories*, *theories of pictorial reference*, and *theories of pictorial expression.*

The content-determining properties that establish what the image depicts can be referred to as the picture's *design*. Thus, a crucial task is to understand how the design properties of a photographic image are mapped onto the domain of meanings, enabling figurative photographs to provide us with the visual experience of a scene composed of objects, with visual properties, arranged in a unified pictorial space. Clarifying how this occurs is the focus of *picture theories*, which we will address in Sect. 2.1. It is not my objective to offer a detailed analysis of the specialized debates in this fruitful field of study. There are different theories dedicated to this theme, and judging their correctness is beyond this book's scope.[2] However, it is important to have at least one way of thinking about how images depict, to formulate the problem of the contextualized use of photographic images, a topic that will be explored later in this chapter.

Nonetheless, the use of photographs extends beyond depiction. Photographs are also employed to *denote* objects, necessitating a distinction between *picture theories* and *theories of pictorial reference*. While picture theories focus on explaining how design is mapped onto pictorial content, theories of pictorial reference are

[1] This distinction between what is depicted and what is communicated can be traced back to the one proposed by Lopes (2004: 160) between "pictorial intention" and "communicative intention." Lopes explores, in pictorial terms, Saul Kripke's (1979) distinction between the semantic meaning of language and the meaning that a speaker attributes to language.

[2] The works of Lopes (2004) and Kulvicki (2006, 2014) serve as invaluable resources for thoroughly exploring these theories.

concerned with how elements of pictorial content are mapped onto the domain of existing objects. Theories of pictorial reference deal with the concept of *denotation*. In addition to examining the different mechanisms of pictorial denotation, our goal will be to analyze the specific case of photography. This will be our focus in Sects. 2.2 and 2.3 respectively.

Moreover, the use of pictures extends beyond depiction and denotation. We also use photographic images to express mental states, feelings, and emotions. The expressive elements of an image play an important role in communication and need to be addressed to achieve an understanding of the ordinary communicative uses of photographs. This is the field of study of *theories of pictorial expression*, which we will analyze in Sect. 2.4. This will later allow us to distinguish between the expressive elements of the fundamental layer of information (encompassing the *utterance act* and the *locutionary act*), the performance of expressive speech acts (a type of *illocutionary act*), and perlocutionary effects.

The study of picture theories and theories of pictorial reference will lead us to question which contextual elements impact and determine what the image depicts and denotes. Understanding these contextual factors will constitute an important cornerstone of photographic communication, which considers communication in relation to the context of use. Just as, in linguistic communication, it is required that the speaker and interlocutor be sufficiently competent in the use of the same language, in pictorial terms, the communicational use of photographs requires that the user and spectator be competent regarding the pictorial system used and be in sufficient agreement about the elements determining the pictorial content. It is by determining a pictorial/propositional content that the image can fulfill the logical role of a locutionary act—as a foundational layer of information linked to an informative intention.

Understanding this is only part of the task to be accomplished in this book, which aims to explain how a photograph (with its specific pictorial content and denotation, contextually determined) can be used by someone to communicate something. In addition to the contextual elements that determine depiction and denotation, we will later seek to understand which contextual elements determine what the image *communicates*. At this point in our analysis (in Chaps. 3 and 4), the idea that we use photographs to perform illocutionary acts becomes central—a second layer indicating that the first layer was intentionally produced and is tied to a communicative intention.

This allows us to recognize that there are two types of contextual elements that impact, in different ways, what the photograph depicts (A) and what the use of the photograph communicates (B). To clarify this distinction, we can apply the division of pragmatic theories of language proposed by Korta and Perry (2011, 2020). According to Korta and Perry (2020), pragmatic theories can be divided into:

(i) *Near-side pragmatics*: concerned with the nature of certain facts that are relevant to determining what is said.

(ii) *Far-side pragmatics*: focused on what happens beyond saying (what speech acts are performed in or by saying what is said, or what implicatures are generated by saying what is said).

Near-side pragmatics focuses primarily on the problem of the ambiguity of signs (a sign can have multiple meanings) and the importance of context for fixing the reference of indexical terms (such as "I," "you," "now," "here," "he," and so on). In cases where there is ambiguity of signs or the need for context to fix the referent, according to Korta and Perry, semantics alone is not sufficient to determine what is said. Understanding contextual facts is necessary for disambiguation or fixing the content of indexical terms.

In a pictorial perspective, the problem of ambiguity arises from the fact that the same design can be mapped onto different pictorial contents, and the same pictorial content can be mapped in various ways onto different domains of existing objects. I will argue that images are inherently *underdetermined* both in terms of *pictorial content* and *denotation*. Therefore, the use of images for the purpose of communication will depend on contextual elements to overcome these underdeterminations. The study of these contextual elements that determine pictorial content and denotation would constitute the field of what I will term *near-side pictorial pragmatics*.

From the perspective of Korta and Perry (2011, 2020), far-side pragmatics is concerned with the study of how we use language in conversational contexts for communication and the actions we perform by using language. In John Austin's terms (1962), far-side pragmatics is dedicated to studying illocutionary acts and their perlocutionary effects. In Paul Grice's terms (1957), far-side pragmatics is concerned with understanding the relationship between communicative intention, the principles and maxims of communication, and the use of implicatures.

From a photographic perspective, the field of far-side pragmatics can be considered in relation to the uses of photographs for communication (which can be distinguished from what they depict). This, in turn, can be considered in relation to the *speech acts* performed through the use of photographs (from the perspective of speech act theory) the contextualized communicative intention, principles, and maxims, as well as the use of implicatures (from the perspective developed by Paul Grice). I will refer to this domain, which is dedicated to understanding speech acts performed through the use of photographs and the role of communicative intention in determining what the use of a photograph communicates, as *far-side pictorial pragmatics*.

Drawing a parallel between pragmatic theories of language and pictures and photographs, Korta and Perry's distinction can be developed as follows:

(i) *Near-side pictorial pragmatics* is concerned with the nature of certain facts that are relevant to determining what is *depicted* (encompassing both pictorial content and denotation).[3]

[3] I will not include here the consideration of contextual elements that determine what an image expresses (mental states, feelings, and emotions), although I will use this idea at certain points in the book. I am omitting it here to avoid adding additional complexity to an already complex topic.

(ii) *Far-side pictorial pragmatics* is focused on what happens beyond depiction (what speech acts are performed in or by using the image, or what implicatures are generated by photographically depicting what is depicted).

It is important to note that it is not my intention to navigate the complexities surrounding the distinction between *pragmatics* and *semantics*—this is a minefield to be avoided. John Kulvicki (2020) considers the contextual facts relevant to determining pictorial content and pictorial reference as part of *pictorial semantics.* I would have no objection to following Kulvicki on this point. Whether termed "near-side pragmatics" or "pictorial semantics," what is crucial for my purpose is that the contextual facts relevant to determining (A) depiction and denotation can be understood as a domain distinct from another that concerns understanding (B) what happens beyond pictorial content and pictorial reference—specifically, what the use of the image *communicates.*

I will explore some of John Kulvicki's ideas, which are part of his *pictorial semantics*, to construct what I term *near-side pictorial pragmatics*, being aware of the terminological disagreement this might cause. Should the reader prefer to label this domain of study "pictorial semantics," this does not impact the fundamental theses I will propose. The domain of pictorial semantics (or near-side pictorial pragmatics) will remain distinct from the field of study related to photographic communication in terms of pictorial speech acts and what the use of photographs communicates (which I describe as part of *far-side pictorial pragmatics*).

Analyzing Kulvicki's position will allow us to notice to what extent the use of pictures can determine propositional contents, introducing the problem of the relationship between pictorial content and propositional content. The idea that the determination of the propositional content of the picture is context-sensitive, in turn, will offer us a way to resolve several problems arising from the parallel between pictures and propositions (which we will explore in Sect. 2.5).

2.1 From Design to Pictorial Content

One of the capacities shared by humans and some animals is the ability to visually segregate content on two-dimensional surfaces (Danto (1992) and Zeimbekis (2015)). Despite their two-dimensional nature, images allow us to perceive colors and shapes as volumetric contents, organized in space. At this stage of visual image processing, we are limited to the segregation processes used by monocular vision and principles such as those described by Gestalt theory, including figure/ground contrast, proximity, similarity, closure, continuity, amodal completion, and illusory contours, among others (Hatfield (2009)). There are also various pre-attentive, sub-personal processes that govern the visual processing of images (Pylyshyn 2003).

This topic will be addressed in Sect. 2.4. For the purposes of this initial outline, the key point to note is how contextual elements determine pictorial content and denotation.

This perception introduces depth, effectively creating a sense of a third dimension. In this case, we see *one spatial arrangement in another* (Wollheim 1987: 62). The space where these volumetric elements are perceived is commonly referred to as *pictorial space.*

However, segregating volumetric contents in two-dimensional surfaces is only the initial stage in picture perception. In the case of figurative pictures (and my analysis will remain primarily restricted to this case), there is an additional requirement: the volumetric contents, once segregated, must be seen as representations of physical objects or their properties. This means that the segregated volumetric contents within the image must be mapped onto a domain of recognizable physical objects.

The elements that constitute the two-dimensional surface can be described, in Robert Hopkins (2015) terms, as *vehicle properties*. The vehicle properties responsible for determining the depicted scene are commonly known as *design properties*. The depicted scene, which includes identifiable objects with visual properties and organized within a three-dimensional pictorial space, is aptly referred to as the *pictorial content.*

The design is the subset of vehicle properties by virtue of which the image represents the depicted scene. In other words, the design properties are the *content-determining properties*. Elements such as brightness, luminosity, texture, reflectivity, contours, hues, light-dark contrasts, etc., constitute the literal properties that typically make up the image's design (Lopes 2004: 3). Not all vehicle properties of a picture are part of its design. The texture of the paper, even though it is a vehicle property, may not play a representational role, such as representing the texture of the skin of a depicted person, although it may play an expressive role, as we will see next. Regarding Hopkins's (2015) terminology, the texture of the paper would be considered a *global content-neutral property*.

Different pictorial theories offer frameworks for understanding how we determine which vehicle properties function as content-determining properties, and how these properties are mapped onto a domain of meanings, enabling the experience of seeing a scene composed of objects within a unified three-dimensional pictorial space. Pictorial theories thus elucidate how we transition from *design* to *pictorial content*. There are at least three types of theories: *perceptualist*, *symbolic*, and *compatibilist*.[4]

Perceptualist theories explain the relationship between a picture's design and the depicted objects by prioritizing perceptual processes. Notable examples include the *resemblance theory* (Peacocke 1987), Gombrich's *illusion theory* (1973, 1977), and Wollheim's *seeing-in theory* (1980). These theories focus on how our perceptual processes determine picture content.

In contrast, symbolic theories approach images as part of symbolic systems, akin to linguistic expressions, graphics, and notations. Images and languages are

[4] Lopes offers this characterization in his 2004 book, along with a detailed critical analysis of the main theoretical approaches.

different types of symbolic systems. From this perspective, what explains pictorial representation is not perceptual in nature (resemblance, illusion, or seeing-in), but rather the syntactic and semantic rules of the system that governs how the characters constituting the image denote. Goodman's theory (1976) and John Kulvicki's structuralist theory (2006) exemplify this approach.

Compatibilist theories attempt to integrate core tenets of both perceptualist and symbolic approaches, seeking a middle ground that acknowledges the strengths of each while avoiding their respective shortcomings and limitations. An example of a compatibilist theory can be found in Lopes (2004).

For comprehensive explorations of these theories, the works of Lopes (2004) and Kulvicki (2006, 2014) are invaluable resources. Given their extensive coverage, I will not go into the specifics of these pictorial theories here. For my purpose, it suffices to note the existence of various approaches and to adopt one of the theories as a basis for understanding depiction. I have opted to base my theory on Dominic Lopes's compatibilist theory. If readers prefer a different pictorial theory, they may simply replace the elements discussed in this chapter with those pertinent to their chosen theory. Nothing central to the communicational theory I will propose hinges on the adoption of a specific picture theory. One can develop an understanding of photographic communication that remains neutral regarding the type of picture theory adopted. (The core of my theory does not fall within the realm of *near-side pictorial pragmatics* but instead lies in the domain of *far-side pictorial pragmatics*). But it is important to understand the conditions that must be met in order to move from design to pictorial content, since it will later be crucial to recognize how this process is context-sensitive.

According to Lopes (2004), pictorial content is what a picture represents the world as having. It is distinct from the physical design of the image and consists in the content that the image conveys when successfully seen by a *suitable perceiver*. This content is determined through the perceptual *recognition* of visual aspects on the two-dimensional surface of the picture, as well as through the grasp of representational conventions embedded in the *pictorial system* employed. Within a certain range of variability (which defines the limits of recognition elasticity beyond which recognition fails), we are capable of recognizing: *i) properties*, *ii) types* of objects, and *iii) individual* objects.

A pictorial system is the set of conventions, styles, and techniques that govern how visual features (such as shape, shading, and spatial arrangement) are used to depict aspects of the world. Different systems, such as Albertian (linear) perspective, X-ray imaging, or anatomical drawing, pick out different aspects of objects and rely on different dimensions of variation and recognition.[5] Competence with respect

[5] An example of the contrast between different pictorial systems in photography can be seen in the distinction between black-and-white and color processes. While black-and-white photography represents the world using luminance, contrast, and texture (excluding color), color photography employs hue, saturation, and brightness to depict additional visual properties. Each system operates under distinct representational conventions and requires different forms of perceptual competence to recognize its pictorial content.

to the pictorial system is essential for identifying which of the vehicle properties functions as content-determining properties.

To recognize the marks on the two-dimensional surface of the image as presenting recognizable aspects of a thing, it is not necessary to possess the concept of the thing or the object. For example, it is not necessary to possess the concept of a "cylinder" to see and recognize one in an image. What is necessary, as argued by Lopes based on David Marr's theory of vision (1982), is the mental representation of the shape in relation to a reference frame based on shape (Lopes 2004: 141). The stable shape encompasses the aspects under which the object can be recognized.

However, in the case of *type* recognition (objects of a certain type recognized as belonging to that type) and *individual* recognition (previously known individuals and objects), a conceptual requirement becomes necessary. I can only recognize a cylindrical shape in a painting as a *mug* if I have the concept of a "mug." Similarly, in the case of recognizing individual objects, concepts are also necessary. I can only recognize the image as that of an individual object or person previously known if I possess the necessary conceptual resources required by that individual content.

Thus, to recognize what a picture depicts (including *properties*, *types* of objects, and *individual* objects), the viewer must satisfy four conditions:

1. Content-recognition: The viewer must recognize the marks on the image's surface as presenting aspects of something—shapes, textures, spatial relations.
2. Subject-recognition: The viewer must recognize those aspects as belonging to a particular kind of object or to a specific individual.
3. Pictorial competence: The viewer must possess knowledge of the pictorial system in use, enabling them to know which vehicle properties are representationally relevant.
4. Conceptual resources: For type and individual recognition, the viewer must have the concepts required to identify the object or person depicted.

These four conditions distinguish *suitable perceivers*—those capable of mapping design onto pictorial content in accordance with the relevant pictorial system—from those who may lack the necessary perceptual or conceptual resources. For instance, viewing X-ray images or infrared thermographs requires specialized pictorial competence; without it, the content of such images remains inaccessible.

Once an observer fulfills these four conditions, they are equipped to recognize the pictorial content of an image, adequately establishing the correlation between *design* and *content*. Lopes refers to images that find such suitable perceivers as *basic pictures* (2004: 152–156). However, not every image is a basic picture. There are images whose content no one is capable of recognizing, preventing anyone from being a suitable perceiver.

The way in which properties, types of objects, and individual objects recognized are combined determines the *commitments* and *non-commitments* of the image. According to Lopes (2004: 112–119), an image is *committal* with respect to a property F if it represents its object as having F or not-F (for example, a man can be represented as having hair or as being bald—without hair). If the image does not provide details about the property, it is *inexplicitly non-committal* regarding that property (such as a photograph of a distant man, which does not allow us to determine whether he has hair or not). Finally, an image is *explicitly non-committal* when it represents its object as having a property that precludes the possibility of it being committed regarding another property (for example, a man represented with a large hat, which prevents the image from indicating whether he has hair or not). For Lopes, the totality of commitments and non-commitments are aspects that, spatially related in a unified whole, constitute the image—"every part of the scene that a picture shows must be represented as standing in certain spatial relations to every other part" (2004: 126). Different pictorial systems will have different explicit and non-explicit commitments.

Lopes's theory is compatibilist in the sense that it integrates core insights from both perceptualist and symbolic accounts of pictorial representation. On the one hand, it retains the perceptualist claim that understanding a picture depends crucially on our visual capacities; particularly our ability to *recognize* properties, objects and individuals. On the other hand, it aligns with the symbolic view in holding that pictorial representation involves *system-relative* rules and that successful picture perception depends on the perceiver's mastery of those rules. In this way, Lopes offers a middle ground: perception is necessary but not sufficient; competence within a pictorial system is also required.

It is important to note the extent to which Lopes's position differs from that of Richard Wollheim (1980). According to Wollheim, for marks on a surface to constitute a picture, two criteria must be met:

A. it must be possible for a viewer to see the object in the image, and
B. the producer of the image must have the *intention* that the object be seen in that manner.

This allows Wollheim to distinguish between seeing an image of X in a certain configuration and that configuration being a *pictorial representation* of X. For example, we might perceive the shape of a rabbit in a cloud in the sky, yet this does not constitute a *pictorial representation* of a rabbit. What the clouds lack is that someone must produce the image with the *intention* that X be seen in that configuration. Lopes refers to this type of intention as "pictorial intention" (2004: 159–160).

Pictorial intention could play an important role in disambiguation. The same arrangement of colors and shapes can be mapped to different pictorial contents (as in the famous case of the ambiguous duck-rabbit image). In this sense, we might say that images are *pictorially underdetermined*.[6] In Wollheim's theory, intention serves

[6] Later, I will distinguish between *pictorial underdetermination* and *denotative underdetermination*, and then distinguish these two cases from *communicational underdetermination*. This will allow us to note that an intentionalist theory of pictorial content (as proposed by Wollheim (1980)), an intentionalist theory of pictorial reference, and an intentionalist theory of communication operate at different levels, so they can be independent in such a way that the adoption of one of these theories does not imply the adoption of the others. Much of the debate about Wollheim's theory (for example, Lopes (2004: 161–167)) does not, in my view, draw a sufficiently clear distinction between an intentionalist theory of pictorial content and an intentionalist theory of pictorial refer-

as the *standard of correctness*, determining which of possible pictorial contents is the content of the picture.

For Lopes, on the other hand, intention need not be a standard of correctness, as the recognition of types, properties, and individuals in the image design is based on processes that are largely subpersonal and automatic. This does not mean that intentionality cannot play the role of a standard of correctness in resolving disambiguation. Lopes's claim is that, in *basic pictures*, the recognition of pictorial content occurs without the need for additional information about the pictorial intention.

Lopes draws an important distinction between *pictorial intention* and *communicative intention* (2004: 160). By pictorial intention, he understands the intention to produce a certain image "embodying information which will enable a suitable viewer to identify something by means of a pictorial mode of identification" (2004: 160). Pictorial intention fulfills the role of Sperber and Wilson's (1995) *informative intention*, insofar as it determines what constitutes the foundational layer of information within the communicative act (as analyzed in Sect. 1.3). Pictorial intention, from Lopes's perspective, is independent of communicative intention, which is what the artist wants to communicate by producing such an image—just as there is a distinction between the *informative intention* (of the image as a fundamental layer of information) and the *communicative intention* of what the user of that image aims to convey. According to Lopes, "artists' intentions are relevant to fixing what artists mean, not what their works mean" (ibid.). By distinguishing pictorial intention from communicative intention, Lopes aims to propose a non-intentionalist approach to *pictorial meaning* (what an image depicts). Lopes's theory, therefore, opposes Richard Wollheim's (1980), in which intention serves as the standard of correctness, determining which of the possible pictorial contents is the content of the picture. But this contrast is achieved through Lopes's focus on the notion of basic pictures, which do not require any criterion of disambiguation.

2.2 From Pictorial Content to Denotation

The use of pictures extends beyond depicting scenes with objects, with visual properties, arranged in a unified three-dimensional pictorial space. Pictures are also employed to *denote* objects in the world, creating a need to distinguish between the study of *picture theories* and *theories of pictorial reference*. While picture theories focus on explaining how design is mapped onto pictorial content, theories of pictorial reference are concerned with how elements of pictorial content are mapped onto the world, in the domain of existing objects. Essentially, theories of pictorial reference deal with the concept of *denotation*.

---

ence (as can be seen in Lopes's discussion of Wollheim's position, where he formulates the problem of picture ambiguity as the problem of determining the referent). In summary, pictorial content, pictorial denotation, and the communicational use of pictures, while connected, are distinct spheres.

The importance of this distinction between pictorial content and denotation in relation to photography can be found in Gareth Evans's work (1982). According to Evans:

> In fact I do not believe that a specification of the content of a photograph should make reference to the object or objects that it is of. [...] We see, here, the need for a distinction between, on the one hand, an *a*-representation (i.e. a species of particular-representation, in a specification of whose content mention of a would figure: something which represents, or misrepresents, *a*), and, on the other, something which, without being an *a*-representation, is a representation *of* a. (1982: 125).

In this case, Evans clearly distinguishes between the *content* of a photograph, constituted by its pictorial content, and its *denotation*, the object or objects that it is *of* (in a causal sense).

Nelson Goodman (1976), a prominent figure in this field, precisely and innovatively demarcated the line between pictorial content and denotation. He posited that an image functions as a label correlating with existing objects. According to Goodman, an image can denote various quantities of objects, ranging from all objects in a specific domain to none. This denotative capacity of an image is theoretically articulated through the linguistic quantifiers "all," "some," "one and only one," and "none."

Take, for instance, the image of a man and a woman on the plaques attached to the Pioneer spacecraft launched in 1972 and 1973.[7] This image was intended to denote *all* human beings. However, its denotative purpose could vary; it could be used to denote *some* humans, *one and only one* particular couple, or, if used as a fictional portrait, it might lack any real-world denotation. In the latter scenario, the image would belong to the type *image-of-man-and-woman*, but would not label any existing man and woman. Goodman distinguished between the label attributed to the image (marked through hyphenation—that determines its type) and the determination of what the image labels, which, in the case of fictional images, would be non-existent, resulting in *null* denotation.

A more complex case (beyond Goodman's analyses) arises when differentiating between denoting *one and only one object* and denoting *a singular object*. For example, an image displayed at the *Alvorada Palace* in Brasília, Brazil, might indicate the office of the current Brazilian president. This human figure denotes the acting president, a role filled by one individual per term, but potentially a different person every 4 years. Here, the image functions as a *definite description*, identifying *one and only one* object at a time, but not a singular, unchanging individual. Contrast this with a picture that denotes a specific individual, like the picture of the president Luiz Inácio Lula da Silva. In this case, the picture's referent is constant over time. Such nuances in pictorial reference theories highlight the complexity and versatility of images as tools for denotation and depiction.

To grasp the concept of denotation in pictures, it is crucial to address the various mechanisms of pictorial reference, which can be formulated as different theories of

[7] This example is put forward by Lopes (2004: 61).

how images refer to objects or entities. These theories include *descriptivism*, *intentionalism*, *causal theory*, and *iconographic* reference. In the context of photography, while causality is often the primary mechanism of reference, understanding how other mechanisms contribute in various communicative settings is essential. The integration of these mechanisms will allow for the establishment of different levels of reference, such as primary and secondary references (I will explain this distinction in the following).

A quite plausible theory to pictorial reference, which will prove problematic and insufficient, is the *descriptivist* approach. Descriptivism hinges on the idea that a description delineates a set of properties, and the denoted entity is the one that fulfills this list of properties. In pictorial terms, a descriptivist approach implies that the picture provides descriptive pictorial content (a list of visual properties). The entities denoted are the objects or object that fulfill the description provided by the picture.

However, the descriptivist approach encounters a significant problem: the same image can be satisfied by numerous visually identical objects. This implies that pictures would be incapable of denoting singular objects in cases where visually identical counterparts exist or could, in principle, exist. Take, for instance, the case of identical twins, Pedro and Paulo. A painting that visually describes Pedro would be equally satisfied by Paulo, making it impossible for the image to denote one twin singularly. In this perspective, pictures are, from a denotative point of view, inherently *underdetermined*, potentially capable of denoting multiple visually identical entities.

It is crucial to note how this *denotative underdetermination* is distinct from *pictorial underdetermination* (discussed in the previous chapter). Pictorial underdetermination pertains to the possibility of mapping the design of an image onto different pictorial contents (as, for example, in the case of the duck-rabbit image). The underdetermination of the denotation pertains to the issue that the same pictorial content can be mapped in different ways in relation to existing objects (the same duck image could be satisfied by numerous visually identical ducks).

In addition to *denotative* underdetermination, another problem with the idea of denotation via descriptive content is expressed by Lopes:

> Conveying true information about people and objects establishes a standard of correctness and usefulness for many pictures, but it is neither a necessary nor a sufficient condition for pictorial reference. (2004: 94)

No object satisfies, for example, the visual description in a caricature or a cubist painting—which distorts the proportions of the depicted subject. Nevertheless, it is possible for an image to have a certain person or object as the referent, even if that person or object does not satisfy the visual properties present in the visual description of the pictorial content. Thus, as Lopes states:

> We must conclude that pictures need not embody descriptive information uniquely true of their subjects in order for viewers to grasp what they denote. (2004: 96)

In the context of handmade pictures, the *producer's intention* becomes a crucial factor in fixing reference, offering a solution to the limitations of descriptivism. It is widely accepted that an image is crafted with a specific intention that determines its referent, regardless of whether other objects might also satisfy the description or whether the visual description does not accurately describe the referent. The artist's intention acts as supplementary information. When coupled with the descriptive pictorial content, it defines the singular object being referred to. For example, if a painting is intended to portray Pedro, one of the identical twins, rather than Paulo, this intention determines that the portrait denotes Pedro, notwithstanding the fact that the visual description could apply to both.

This concept of intention as a determinant in pictorial reference can be understood in relation to the distinction proposed by Keith Donnellan (1966) and further developed by David Kaplan (1978, 1989) between *descriptive* (termed "attributive" by Donnellan) and *referential* uses of descriptions. In descriptive usage, the referred entity is any object that satisfies the descriptive content. However, descriptions are often used *referentially* to indicate a specific person, even if they do not fit the description. In such instances, the speaker's *referential intention* fixes the reference. When this intention successfully identifies the referent, even if the description is factually incorrect, the object itself becomes part of the propositional content of the statement made. For instance, if someone says "the young man drinking *cachaça* leaning against the door is a professor," it can successfully refer to a particular person even if the individual is actually drinking water, not *cachaça*. From a satisfactional or attributive perspective, the description would fail to denote that person. In David Kaplan's framework, the referential use of a description *rigidifies* it, akin to placing a demonstrative before the description. When employing the sentence referentially, the expressed proposition is: "[that person]—whom we identify by saying 'the young man drinking *cachaça* leaning against the door'—has the property of being a professor." In this scenario, the definite description is not part of the singular proposition, which would be reformulated as: "[that actual person] has the property of being a professor."

Similarly, in the realm of pictures, intention can shift the use from descriptive to referential, thereby *rigidifying* the pictorial content. This approach addresses the two main issues identified in descriptivism. A rigidified visual description can denote a singular object even if other objects also fulfill the description. It can also denote an object even if the description contains inaccurate visual information about the referent. This approach to picture reference emphasizes the significance of the producer's or user's intention in determining the referent, going beyond the confines of visual description.

Iconographic reference presents yet another mechanism through which images can denote objects. This is one of the cases of "non-depictive content" described by Robert Hopkins (Sect. 1.2), when someone depicts one thing with the intention of symbolizing another—for example, "the dove and peace" (2015: 345). Iconographic reference uses elements that are conventionally associated with objects, turning parts of the pictorial content into *symbols* that conventionally denote singular entities or groups of entities. The elements of the picture thus function as emblems that denote particular objects by convention.

In the realm of iconography, especially in religious or cultural contexts, certain symbols and emblems are recognized and understood to represent particular figures or concepts. For instance, in Christian iconography, an anchor is commonly associated with Saint Clement, a figure with tied arms and pierced by arrows denotes Saint Sebastian, and a tray with eyes refers to Saint Lucy. These iconographic elements are not mere visual descriptions of their subjects; they become *emblems* that conventionally denote these singular entities. These would be cases in which denotation via descriptive content would not be possible, as there is no *suitable spectator* for these pictures, since no one knows what Saint Clement, Saint Sebastian, or Saint Lucy actually looked like to recognize them in the images.

In the realm of photography, causality emerges as a central mechanism of reference. We generally use photographs in such a way that, if Pedro, for instance, posed for the photograph, being the causal referent, the image denotes Pedro and not Paulo. The image denotes only one of the twins due to its causal link to that object.

Similar to intention, causality can serve as a means to *rigidify* the descriptive content of an image. This allows the image to have a singular object as its denotation, even if the pictorial content is not an accurate portrayal of the referent or is satisfied by multiple objects. Thus, an image can be a picture *of* something, in the causal sense, even if its descriptive content does not accurately depict that something. The causal relation selects the object in the world that is the causal referent of the photograph and connects it to the pictorial content in a non-satisfactional way.

Kaplan describes this causal mechanism of fixing reference as *metasemantic*, as it operates prior to the determination of the semantic content of our thoughts about the photographed entity during picture perception (1984: 574–576). Even if I do not know which of the twins I am thinking about when looking at the photograph, before the determination of the content of this thought, causality already fixed (*externally,* and logically prior to my thought—and, in this sense, *metasemantically*) which of the brothers would be the referent of the picture. This would be a form of *semantic externalism*, because, if I do not know who the referent is, the reference would be fixed externally to my thought by the causal relation. In this case, we would have to abandon "Russell's Principle," as expounded by Gareth Evans (1982), which states that in order to have a singular thought (about a singular object), we must know which entity we are thinking about. (I will return to this point later). The *externalist* aspect is related to how what determines the content of the thought (namely, causality) would have the capacity to fix an object as content, even if we do not have *in* our thought the determination of which object we are thinking about.

In the case of handmade images, it is widely acknowledged that intention is often a more decisive factor than the causal relation. For instance, if Paulo had passed away, using Pedro as a model to create a portrait intended to represent Paulo would be generally accepted. In photography, the role of causality as a mechanism for fixing reference is generally considered to outweigh intention.

2.3 Different Modes of Photographic Reference

Neither a purely causal approach nor a purely descriptive approach provides a fully adequate understanding of photographic pictorial reference. In the case of photography, we want not only the image to be causally connected to the denoted object but also the pictorial content to visually describe the referent in a way that enables its recognition. Nelson Goodman provides the following anecdote, which explores this issue:

> If I tell you I have a picture of a certain black horse, and then I produce a snapshot in which he has come out a light speck in the distance, you can hardly convict me of lying; but you may well feel that I misled you. (Goodman 1976: 59–60)

Since it is a common belief that causality is the primary mechanism in photography for establishing reference, we are inclined to accept that an image denotes a singular horse even if it does not accurately describe that horse. Causality can *rigidify* the pictorial descriptive content. However, there is a threshold to the permissible inaccuracies in the image's descriptive content, beyond which we would feel misled, as Goodman illustrates. We do not only want *causal connection* but also *depiction.*

To understand pictorial reference in photographs more effectively, it is essential to move beyond the Orthodox view that understands the connection between photography and the world as centered on the causal mechanism. According to the view that should be avoided, the way a photograph denotes would always be determined by its nature (as part of metaphysical considerations), by how the image is constituted through a causal relation (in a mind/belief-independent way). Instead, according to the theory I propose, we should adopt an interlocutory perspective and recognize that our *usage* in specific contexts determines the mechanism of pictorial reference.

In contexts where social practice rules the use and production of photographs, enabling it to function as a reliable informational system, there is an underlying assumption that the pictorial content of a photograph provides visual information facilitating the recognition of its causal referent. The image is not only accurate (or truthful) but also genuine. In such cases, there is a congruence between the causal genesis and the pictorial descriptive content. We can refer to this scenario as *basic photographic pictorial reference*, adapting the terminology from Lopes (2004) to photography. But there are various non-basic uses of pictorial reference. Let us consider examples where photography is used differently and how this impacts the relationship between photography and the world.

Take the scenario of choosing a sandwich from a photograph on a menu. In this context, if we point to a photograph and say, "I want *this* sandwich," it would be a communicational error if the attendant brought the actual sandwich that is the causal source used in the photoshoot; which, in food advertising, is often an inedible mockup. Here, the demonstrative "this" does not refer to the causal referent of the photograph. Instead, it refers to the visual properties displayed in the photograph, exemplifying the characteristics an object should have to qualify as the requested sandwich.

This use shares important parallels with Zeimbekis' concept of *pictorial indexical predication* (2010), which is based on Jane Heal's indexical predication (1997). When we say, while pointing to a photograph, "this is so-and-so," according to Zeimbekis, the demonstrative "this" is not used to identify a singular object, but to point to the visual perceptual properties exemplified by the photograph, which are part of the predicate *attributed* to the referent. This position stems from the acceptance of "Russell's Principle" (Evans 1982), which states that in order to have a singular thought (about a singular object), we must know which entity we are thinking about. Zeimbekis argues that picture perception would not suffice for knowing which object we are thinking about. For example, if we do not know the causal history of a photograph (how it was produced and which object was part of the pro-photographic scene), when confronted with a photograph of one of the twin brothers, we would not know which brother we are thinking about. The same applies to photographs in general, since, from a descriptive point of view, their content could be satisfied by numerous visually identical objects. Thus, according to Zeimbekis: "the contents of picture perceptions do not themselves provide the kind of numerical and contextual information required for singular thought" (2010: 11).

In his argument, Zeimbekis explores an important difference between the perception of objects and the perception via photographs. If we were in front of one of the two twin brothers, even if we did not know whether it was Pedro or Paulo, we could locate in space and time the singular object to which we are referring with the demonstrative "this" (or "that"). As Evans (1982: 143–204) and Strawson (2003: 19–20) formulate, we can *demonstratively identify the object*, and thus, we would have the ability to distinguish the object in thought (fulfilling the requirement imposed by Russell's Principle). We are thinking about "*this*" object, spatially and temporally located through perception. We would thus have a "singular thought based on perception" (Zeimbekis 2010: 15). The determination of the referent would not occur through an identification based on the visual properties of the object (in a descriptivist manner), but by locating the referent in space and time (within the egocentric frame of reference, within the restricted domain of perception). Photographs, on the other hand, are *spatially* and *temporally agnostic*—they do not allow us to determine the location of the depicted object in relation to egocentric space and time.[8] We can locate the photograph as a material object (a printed sheet, for example) in egocentric space, but not the objects depicted in the image. As Mohan Matthen explains, the space in the picture lacks a "here," and similarly, the time in the picture also lacks a "now" (2005: 316–317).

Let's suppose I am the producer of the photograph and know which of the brothers is the causal referent. In this case, knowing its causal history, I can, upon looking at the photograph later, have a singular thought about that particular object (because I know which object was photographed). What enables me to have the singular thought in this case is not the pictorial content of the photograph or the existence of

[8] The idea of photography as "spatially agnostic" is explored by Aaron Meskin and Jonathan Cohen (2008).

a causal route that links the photograph to the referent, but rather my prior knowledge of the referent. In other words, if I point to the photograph and say "*this* is so-and-so," the determination of the referential part of the sentence (which singular object I am thinking about) would not be provided by the pictorial content but by knowledge external to the photograph (my direct knowledge of the referent that was photographed). Due to this type of underdetermination, if restricted to pictorial content without additional information, when using photographs in demonstrative contexts, according to Zeimbekis, we would be incapable of singling out the referent.

Even if we do not know which object we are thinking about, we can still point to a photograph and say "*this* is so-and-so." But what would be the logical function of the demonstrative "this," given the incapacity of the demonstrative act to singularly identify the referent—in a way that aligns with Russell's Principle? Zeimbekis' answer is that, when one points to a photograph and says "this is so-and-so," the intended meaning is "the so-and-so is like *this*"—and the demonstrative points to the visual properties exemplified by the photograph (2010: 19). The photograph does not give us the referent but the visual properties exemplified by the image that are *attributed* to the referent. For this reason, it would be a form of *indexical predication* (as outlined by Jane Heal (1997)).

Similarly, when someone says, "I want *this* sandwich," they are expressing the thought "I want a sandwich that looks like *this*." The photograph serves as a model, setting the visual standards for a particular category of sandwich.

But here we have a complete opposition between Zeimbekis' and Kaplan's understandings regarding causality and the acceptance or rejection of Russell's Principle. Zeimbekis adopts a form of *semantic internalism* that requires us to know which object we are thinking about for singular thought. Given that the descriptive content of a photograph is underdetermined, even if the photograph is causally linked to the object, we would not know which object we are thinking about. Thus, although we may believe a causal route links the photograph as an object to the photographed object, causality would be insufficient for determining the singular referent. Kaplan's position adopts a form of *semantic externalism* that abandons Russell's Principle and accepts causality as sufficient for determining the referent, even if we do not know which object we are thinking about. There are thus two antagonistic approaches that treat the causal relationship between the photograph and the depicted object in distinct ways:

- *Semantic Externalism applied to photographs*: Causality is sufficient to determine the content as singular.
- *Semantic Internalism applied to photographs*: Causality is insufficient to determine the content as singular.

The fundamental point of this disagreement is the acceptance or rejection of Russell's Principle. The critical question then becomes: should we accept the restriction of Russell's Principle on singular thought and, consequently, that photographs do not provide the information required for singular thought? Or should we abandon Russell's Principle and take causality as sufficient for determining singular content?

In philosophy, we often accept too hastily that the existence of antagonistic theories implies that one must be correct and the other wrong. What I would like to suggest is that by prioritizing the understanding of the *use* of photography in communicational contexts, the fundamental question should not be "which theory is correct?" (where the answer would be sought in relation to our intuitions about the essence of photography) but "which theory can explain a certain contextualized use of a photograph?". My position is that there are contexts of use and communicative intentions in which a semantic externalist perspective on reference provides a better framework for understanding that use. Conversely, there are also contexts of use and communicative intentions where an internalist perspective on singular content is more appropriate. In the case of the sandwich menu, for example, Zeimbekis' approach is preferable, especially given the fact that I am unable to have a singular thought about the *future* sandwich I intend to eat (since I have no acquaintance with it).

However, in relation to the menu of sandwiches, the point here is not to deny that the photograph has a causal referent. The key point is that in this *photographic social practice*, there is the *rule* that it is not the causal relationship that fixes the *primary referent* of the demonstrative act (expressed in language when we say "*this* sandwich"). The primary referent is *any* sandwich that satisfies the visual properties depicted in the photograph used as a model.[9] The singular causal referent (used in the photoshoot) can be considered a *secondary reference*. In this terminology, the distinction between primary and secondary reference hinges on what is primarily targeted in the actual contextualized use.[10] In the sandwich menu example, distinguishing between the primary and secondary referents of the photographic image depends on how we *prioritize* the pictorial descriptive dimension (treating the photograph as a model) versus its causal mechanism in fixing the referent.

An extension of this example is found in cases where a photograph is labeled "for illustrative purposes only." Here, the photograph is not used to determine the specific visual properties required for an object to belong to a particular class (a certain type of sandwich) but merely illustrates a general class (of sandwiches).[11]

A similar situation occurs in pictorial dictionaries, where photographs often accompany entries to illustrate the class of the object defined. While these photographs have singular content (fixed by their causal relationship with the singular photographed object), they are not meant to denote only the causal referent. Instead,

[9] In Chap. 3, we will address how, in cases like this, the *direction of fit* of the image is inverted compared to its basic referential assertive use. It is not the image that should correspond to the world, but the world should correspond to the image as a model.

[10] Lopes proposed a similar distinction between primary and secondary sources (2004: 164).

[11] In Chap. 3, we will note how the use of the caption "for illustrative purposes only" weakens the *commissive illocutionary act* performed through this use. Besides the *directive* act of making the viewer buy the product, the advertising use of photographs also involves a *commissive* act, in which if the viewer buys the product, the seller must provide something similar to the descriptive content of the photograph. The caption "for illustrative purposes only" restricts the commissive act to the category, without requiring the sold object to satisfy the visual description.

they are used to denote all objects within the class to which the causal referent belongs. Therefore, a dictionary may use a photograph of a singular dog to refer to the entire class of dogs, even though other breeds may not share the visual properties of the class exemplified in the image. We could also have other levels of generality in which a photograph of a dog could be used to denote the class of animals, or mammals, etc. This is the second case of "non-depictive content" described by Hopkins, as discussed in Sect. 1.2, where a photograph might depict a particular object to exemplify a *type* (Hopkins 2015: 345).

The pictorial content of a photograph can also denote through *iconographic* reference. In such cases, a part of the image is transformed into a symbol that is conventionally associated with a referent and can denote varying quantities of objects. This process of converting a segment of the image into a symbol that denotes iconographically does not eliminate the causal and descriptive dimensions of the picture, but rather complements them, creating multiple levels of reference such as primary, secondary, tertiary, and so forth.

For example, consider a photograph where the primary reference is determined by a causal mechanism and the secondary by iconographic reference. A photograph with an Urubu bird as its causal referent, that depicts an Urubu, can be used as a symbol that, by iconographic convention, represents a specific group of people. In Brazilian football iconography, an Urubu denotes the supporters of the Flamengo team. In this scenario, the singular Urubu could serve as the primary referent (determined both causally and descriptively), while the group of Flamengo supporters denoted by iconographic convention would function as the secondary referent—or vice versa. The visual and expressive properties of the image could then be attributed to the group it iconographically represents. A photograph of a fierce Urubu, for instance, might be used to associate qualities such as boldness, resilience, or defiance with Flamengo fans. This process, starting from a singular case and extending to general predication, could be termed *pictorial generalization*.[12] Thus, generalization may occur either iconographically, when an image is used to represent a group or category by established visual convention, or descriptively, when a particular case is employed to exemplify a type.

However, the prevalence of Orthodox intuition often leads us to assume that singular causal reference is the exclusive mode of reference in photographs.[13] Orthodox theories tend to prioritize the causal mechanism for fixing photographic reference and either exclude or greatly diminish the potential for other forms of pictorial reference within photography at other levels of generality. In reality, there are situations where causal reference is set aside (as exemplified by the sandwich menu example) or used to express a content that is not singular (by pictorial generalization or iconographic reference). Consequently, even when causality serves as a

[12] As a consequence of pictorial generalization, as I will argue later, the speech act performed by the photograph (assertive, expressive, etc.) may also have different levels of generality, depending on how it is used and intended in the context. Thus, a derogatory photograph of a person can be used as a general critique (of all persons from a certain place, race, gender, etc.).

[13] Although this is not a mistake committed by Robert Hopkins (2015).

mechanism for establishing primary reference, this does not preclude the same image from concurrently supporting other levels of reference.

Moreover, contextualized use may require an understanding of which domain of quantification is being considered, whether an unrestricted domain or a specific domain. For example, for a photograph that has a particular actor as its causal referent to be used in such a way that, in context, the primary reference is the fictional character Ulysses, the interlocutors must understand that the use of the image pertains to the domain of quantification over fictional characters from Homer's Odyssey.

2.4 Theories of Pictorial Expression

Photographs are not only used to depict scenes or denote objects but also as a means of expressing mental states, feelings, and emotions. It is common to describe photographs as being cheerful, sad, strong, evocative, poignant, dramatic, serene, haunting, vibrant, and so on, illustrating diverse ways in which a photograph can convey expressiveness. Since expressiveness can play a crucial role in communication through photographs, it is important to understand the various ways an image can be expressive. This understanding will later allow us to distinguish three dimensions of expressiveness: *i*) the expressiveness of the picture itself, *ii*) the use of the picture to perform an expressive illocutionary act, and *iii*) the perlocutionary effects that result from the use of the image.

A significant contribution to the study of pictorial expression is found in Dominic Lopes' book *Sight and Sensibility: Evaluating Pictures*, published in 2005. I will use Lopes' ideas as a foundation for the reflections in this section and explore some of the themes he introduced.

The most straightforward way for an image to be expressive is by depicting a human or animal figure, or an anthropomorphized object, whose characteristics—such as facial expressions or body positions—convey a particular mental state, feeling, or emotion. Many mental states have characteristic behaviors, and by portraying bodies that exhibit these behaviors, an image can express pain, joy, sadness, pleasure, and so on. Let us refer to this type of expressiveness as *figure expression* (Lopes 2005: 51).

However, an image can also be expressive even without the portrayal of a human figure, an animal, or an anthropomorphized object. In such cases, the expressiveness of the *scene* depicted, or a part of it, cannot be attributed to the *figure*'s expression. For instance, a landscape can be contemplative, a still life can be nostalgic, and a stormy sea can be threatening. In these instances, there may be no figure whose behavior expresses the mental state conveyed by the image.

Furthermore, even when an image includes a figure expressing a certain mental state, the expressiveness of the scene may not be reducible to the expression of the figure. There are situations where figure expression aligns with scene expression, conveying the same mental state. However, it is also possible for there to be a disconnection between figure and scene expression. For example, a sad human figure can be part of a comic scene, just as a smiling figure can be part of a sad scene. Therefore, *figure* expression and *scene* expression should be considered distinct and explained independently, without one being reduced to the other.

Additionally, there exists a third mode of expression. As emphasized by Lopes (2005: 55–57), figure expression and scene expression depend on what the image depicts, making them modes of expression related to the *pictorial content*. However, the expressive content of the picture is not confined to its representational aspects; it can also be linked to its *design*, encompassing elements by virtue of which the picture represents (such as brightness, luminosity, texture, reflectivity, contours, hues, and light-dark contrasts) (Lopes 2005: 57).

To understand how design can be expressive independently of the pictorial content, we can focus on abstract images. Abstract images lack discernible figures or recognizable scenes, so their expressiveness cannot be reduced to the previous two modes. Nonetheless, abstract images can express dynamism, joy, strength, sadness, contemplation, and so on. The same elements that underlie the expressive content of abstract images are at play in pictorial images, constituting the expressiveness of the picture's *design*.

Design expression constitutes the third mode of expression and can either align or diverge from the mental states, feelings, and emotions expressed by the figure and the scene. For instance, a sad figure in a sad scene may be depicted using cheerful colors, just as a joyful figure and scene can be rendered with somber colors.

In the realm of human expressions, Dominic Lopes categorizes expressions as *natural*, *conventional*, or *sui generis* (2005: 59–60). There are facial and body configurations that are *naturally* associated with specific feelings and emotions. These stereotypical expressions are "innate psychophysiological mechanisms" that link behavior with mental states (Lopes 2005: 60). Even though it is possible to perform such actions without genuinely experiencing the corresponding feeling (as in the case of actors pretending to feel pain), these behaviors are considered *natural* expressions of their respective mental states.

Conventional modes of expression also exist, where certain gestures or behaviors conventionally signify specific emotions. For example, contentment can be conventionally expressed by raising the hand with the thumb up, while discontent can be expressed by making a similar gesture with the thumb down. Different eras, cultures, and contexts may have distinct repertoires of conventional expressions.

Sui generis expressions are cases where the expression is neither natural nor determined by convention. Lopes provides examples of *sui generis* expressions, such as expressing anger by keying a car, expressing joy by sliding down a banister, or expressing sadness by sitting alone on a beach (2005: 60). While there may be no convention dictating that sliding down a banister signifies happiness (as in the case of the thumbs-up gesture), such actions can still be expressions of that specific emotion in particular contexts.

However, the challenge arises when we attempt to explain scene expression and design expression, since there may be no human or animal figure in the image that serves as the basis for its expressive content. Lopes terms this challenge the "missing person problem" (2005: 57–58) and offers four approaches to address it:

1. In *personalism*, the mental state, feeling, or emotion expressed by an image is attributed to the creator of the image. However, personalism faces limitations as it does not account for situations where the feeling expressed by the image may differ from what the creator feels. For instance, a sad image could be crafted by someone who is experiencing happiness, another emotion, or even no emotion whatsoever.
2. *Hypothetical personalism* proposes that the emotional state expressed by the scene or design is attributed to someone related to the image's context but who is not necessarily portrayed in the image. This person may be entirely hypothetical, and the emotional attribution is based on contextual factors.
3. *Arousalism*, on the other hand, posits that the bearer of the mental state, feeling, or emotion expressed by the image is the viewer. However, this approach encounters challenges, as the feeling experienced by the viewer may not always align with the feeling expressed by the image. For instance, a sad image might evoke pity in the viewer, but the image expresses sadness, not pity.
4. To address these difficulties, Lopes introduces the idea of an *impersonal theory of pictorial expression*—dubbed by him as the "contour theory" (2005: 72). In this theory, a figure, scene, or design expresses a certain feeling or emotion if the expression-look is a configuration that, in certain circumstances, serves to indicate the emotion. This association between *appearance-of-expression* and emotion can be natural, conventional, or *sui generis*, and it depends on the relationship the image has with its context.

I believe it is not necessary for the purposes of this book to exclusively choose one approach as the correct one. Lopes acknowledges the importance of *practices* and *contexts*, suggesting that different contexts may warrant attributing the expressed emotion to different bearers (the creator, a hypothetical subject, the viewer), or adopting an impersonal stand. The choice of attribution can depend on the need to make sense of a narrative or artistic purpose, among other factors.

In my usage of Lopes's theory, I will propose initially two additions. First, we should focus on the picture's *user* rather than its *producer*. This perspective allows for a *personalist* or *hypothetical personalist* approach where the emotion an image conveys is attributed to its user, even if they are not the creator. This scenario is typical in photographic communication, in which people often use photographs they did not create to communicate how they feel.

Second, we can expand Lopes's characterization of the expressiveness of *design* to include the expressiveness of the picture's *vehicle properties* that do not play a representational role. As seen in Sect. 1.1, Hopkins (2015) explores this expressiveness by analyzing photographic communication in relation to *content-neutral* properties, which can be either *global* or *local*. This consideration can be set aside when focusing on the expressiveness of a *picture*, as the content-neutral properties are not part of picture's *design*. However, especially in artistic contexts, attention to the expressiveness of the material support of the image plays an important role. Different materials can significantly influence the aesthetic experience of a photograph, as they affect its overall expressiveness (as in the example explored by Robert Hopkins

(2015: 341–342) of Shomei Tomatsu's photograph "Beer Bottle after the Atomic Bombing", in which the expressiveness is related to a content-neutral global property).

In summary, for our purposes, it is important to note that in different uses of photographs, in different contexts, the expressive content of the image may be related to:

(i) Figure expression (human, animal, or anthropomorphized objects);
(ii) Scene expression (the pictorial content as a whole);
(iii) Design expression (colors and shapes by which the pictorial content is determined);
(iv) Content-neutral vehicle properties expression (the material support of the picture).

These can be called the four *modes of pictorial expression.*

The expressed content can be conceived as a mental state, feeling, or emotion:

(i) From the creator/user of the image;
(ii) From a hypothetical subject not portrayed;
(iii) From the viewer;
(iv) Or just an expression-look without a bearer.

These can be called the three *foci of attribution of pictorial expression.*

Finally, the content can be expressed by being associated with a particular emotion in a way:

(i) Natural;
(ii) Conventional;
(iii) Or *sui generis.*

These can be called the three *mechanisms of pictorial expression.*

The complexity of the expressive content of a picture arises from the fact that various modes of pictorial expression can operate simultaneously, amplifying what the picture conveys. In such instances, the expressiveness of the figure, scene, design, and vehicle properties may all be active within the image at the same time. Modes of expression can converge to convey a particular mental state, feeling, or emotion. It is also possible for these modes to diverge, resulting in a highly complex expressive content. An image might feature a sad figure in a tranquil scene, with a design that conveys liveliness, using vehicle properties that evoke nostalgia.

Furthermore, different *foci of attribution* of pictorial expression can coexist within the same image. A user who presents a photograph may, through the use of that picture, express their own anguish with the intention of eliciting pity from the interlocutor, by using an expression-look of a figure associated with sadness.[14]

[14] In Sect. 3.5.4, we will distinguish between the expressiveness of the picture (which includes the cases analyzed in this chapter), the realization of expressive illocutionary acts, and the perlocutionary effect. In relation to the theory I will propose, we could analyze this example as the action of a

What enables us to incorporate these ideas on picture expression within the horizon of a *near-side pictorial pragmatics* is the fact that many of these elements are *context-sensitive* and can be related to what the image *depicts*. What an image depicts may include the expressive elements, as these expressive elements can be taken as *properties* of what is *depicted*. Thus, expressiveness can be considered part of the pictorial content. In a photograph where a human figure expresses joy in a scene that expresses sadness, the pictorial content of the image would be that of a joyful person in a sad scene.

When someone uses a photograph in an interlocutory context, as part of an intentional ostensive act of communicating something to someone, the expressive elements of the picture can be recruited as part of what is communicated. Thus, for communication to be possible, in these cases, it is necessary for the user and the viewer to agree on which expressive elements are recruited by the user and should be grasped by the viewer, so the communicative intention is recognized. This may include understanding which *modes*, *foci of attribution*, and *mechanisms* of pictorial expression are used. For example, if the user intends to perform a derogatory act regarding the person portrayed, using a photograph in which this figure expresses comicality in a *sui generis* manner within a sad scene, the communication will only be possible if the viewer can recognize the *sui generis* behavior as expressing comicality within the sad scene.

Additionally, there is another level of complexity in the use of expressive elements. Communication that recruits the expressive elements of an image *may* or *may not* require the viewer's recognition of the focus of attribution of pictorial expression. This will allow us to distinguish between Gricean and non-Gricean uses of expressions as part of photographic communication. In Paul Grice's theory (1957), effective communication requires the speaker's communicative intention to be manifest and recognized by the interlocutor. For instance, for a verbal statement to successfully order an interlocutor, the latter must recognize the speaker's intent to issue an order. Fodor (2012), in his analysis of Arthur Danto's theory of art (1981), differentiates between uses of images that depend on *reflexive intention* and those that do not. For instance, in the use of a photograph for blackmail, it is crucial for the viewer to recognize that they are being blackmailed through the use of that picture.[15] This recognition of the user's intention is what constitutes the use of the image as an act of blackmail, qualifying it as a Gricean case where the photograph serves as a vehicle of communication under *reflexive conditions* (Fodor 2012: 62). (If the viewer did not recognize that they were being blackmailed, the directive act of coercing the viewer to take a certain action under the threat of retaliation would not be properly implemented. We will address these points in detail in the following chapters).

---

user performing an expressive speech act (expressing their own anguish) to elicit a perlocutionary effect in the interlocutor (eliciting pity), using a picture that expresses sadness.

[15] We addressed Fodor's notion of a reflexive intention in Chap. 1.

In contrast, there are non-Gricean uses of photographs where reflexive intention is not central or may even be intentionally obscured. A photograph can act as a guide, directing the viewer to perform certain actions without recognizing the user's intent. Fodor notes an example of an advertisement that influences the viewer to buy a product in a "subliminal and associative" manner (2012: 63). The image thus harbors intent to induce the viewer to purchase a product, achieving its aim through "hidden persuasion" (ibid).

Thus, we can characterize *Gricean cases* of the use of the expressiveness of a picture as those that require the viewer's recognition that a certain expressive element is part of the user's communicative intention. We may also conceive as *Gricean cases* those in which the use of the expressiveness of the picture with the purpose of performing an expressive act to be recognized by the viewer is part of the function of the social photographic practice. In many artistic practices, for example, the purpose and value of photographs are related to how the expressive possibilities of the picture are intentionally used as part of a communicative intention to be recognized by the viewers within the practice (we will explore this idea further in Sect. 3.5.4). This can be distinguished from *non-Gricean cases* where expressiveness functions as part of hidden persuasion in a subliminal and associative manner. For example, it is possible for a user to employ a sad image to evoke remorse in the viewer (in line with an *Arousalist* approach to pictorial expression) but in such a way that the remorse works subliminally.

But for the purposes of this chapter, the fundamental point is that there are context-sensitive elements that determine what an image expresses and how it expresses it. These context-sensitive expressive elements can be recruited in communication as part of what a picture depicts, such that the user and the viewer must share a common understanding for communication to be possible. This could be circumscribed as part of *near-side pictorial pragmatics* (related to the contextual determination of pictorial content). Later, we can contrast this with the idea of expressive illocutionary acts, which could be subsumed as part of f*ar-side pictorial pragmatics*.

2.5 Context Sensitivity of Pictorial and Propositional Content

A key point for understanding the use of photography as a means of communication is to note that both the determination of pictorial content and denotation are *context-sensitive* processes. The transition from design to pictorial content depends on competence regarding the pictorial system used and the conceptual framework applied in mapping the design onto the domain of meanings. Since these elements are context-sensitive, the pictorial content of an image is also be context-sensitive. Moreover, the mapping of pictorial content onto the domain of existing objects could employ different mechanisms of pictorial reference and exhibit varying levels of generality, also in a context-sensitive way. These context-sensitive elements need to be shared by both the user and the viewer for an image to be effectively used as a means of communication.

These ideas can be addressed here in the following terms: the *propositional* content determined by the use of a picture as the foundational layer of information is *context-sensitive*. However, the parallel between pictures and propositions—which allows us to conceive of the use of a picture as the execution of a locutionary act—is not without its problems, as we shall now see. In this chapter, I will employ the parallel that Kulvicki (2020) draws between the use of images and David Kaplan's (1978, 1989) theory of indexicals to explore these ideas.

Propositions are linguistic constructs that serve as bearers of truth-value. A proposition represents a complete thought, capable of being classified as either true or false, depending on whether the conditions it posits are satisfied or not. An argument advanced by Alex Grzankowski (2015) presents a compelling case for the view that pictorial content is, to some extent, propositional. He argues that we can apply linguistic tools such as *modal operators* and *negation* to pictorial content. For example, when one refers to a photograph of a man leaping over a car and asserts "this is possible," the use of a modal operator (possible, impossible, necessary, contingent) to the picture's content suggests its propositional nature. Since modal operators operate on propositional content, the pictorial content must be propositional. Similarly, the ability to negate a picture's content by stating "this is not what happened" underlines the propositional aspect of pictorial content. What is negated must be a proposition. Grzankowski's argument could be further extended to encompass tense operators, such as "it was the case that," "it is the case that," and "it will be the case that," all of which also imply that the content they modify is propositional.

However, the correlation between pictorial content and propositions is not as direct as the relationship between the signs of a language and the propositions we express through the use of those signs, presenting a challenge to the idea that pictorial contents can be used to express propositions. It is the structured use of language, following syntactic rules, that enables the creation of sentences expressing propositions, each carrying the semantic significance of being either true or false. But, unlike language, pictures lack a well-defined syntax for determining the propositional content they convey. In verbal language, the subject-predicate structure offers clarity regarding the referent and the properties attributed. Languages like German, Greek, and Latin further enhance this clarity through their use of declensions to determine syntactic functions. Pictures, in contrast, do not offer such syntactic demarcations, leaving open the question of what is being referred to and what properties are being ascribed to which object within the pictorial content. This lack of syntactic specificity in pictures leads to an inherent ambiguity in interpreting pictorial content in relation to propositional content.

One interesting and fruitful way to understand the extent to which pictures may express propositions, and to circumvent the syntactic ambiguity of pictures, can be found in John Kulvicki's (2020) use of David Kaplan's theory of indexicals. This will allow us to deepen the idea that the relation between pictorial and propositional content is context sensitive. Furthermore, we will be able to notice precisely how it is the *use* of the picture that determines propositional content. This enables us to recognize the importance of communication occurring in contexts where users present pictures to viewers with the intention of communicating something.

According to David Kaplan (1978, 1989), there are elements of language whose contents vary from context to context and elements that are insensitive to contextual variation. The theory focuses on indexicals, which include terms like "I," "here," "now," "he," "this," etc., whose references vary according to the context of use. In Kaplan's theory, context is associated with an *agent*, *time*, *place*, and *possible world*. The content of "I" in the sentence "I am an artist" varies if the utterance is made by different agents (speakers) in different contexts. When the sentence is uttered by Pedro, the referent of "I" in that context is Pedro, and the sentence expresses the singular proposition that attributes to Pedro the property of being an artist. In another context, uttered by another speaker, the sentence would express a different proposition, and the indexical term "I" would have a different referent as its content. If, at the relevant time of the proposition's evaluation, Pedro is an artist, the proposition will be true. The time at which the content and truth value are determined is important because there could have been a time when Pedro was not an artist (in his childhood), and the sentence he uttered would have been false. Additionally, in a possible world where Pedro is a monk and not an artist, the truth value of the sentence he utters would also be false.

Kaplan's seminal contribution to the understanding of indexicals is the development of the distinction between *character* and *content*. The character is a rule (a *function* in the mathematical sense) that, in different contexts, determines different contents. More precisely, the character is the rule by which the indexical is mapped onto the domain of contents. The indexical "now" has as its character something akin to "the time simultaneous with the utterance." The indexical "I" has as its character "the one who speaks at this moment." Although the rule present in the character of the expression is constant, in different contexts, the expression may have different contents. The content, in turn, is the referent of the term.

The parallel proposed by Kulvicki between images and Kaplan's theory stems from the distinction, already present in Kulvicki's 2006 book, between the *bare bones content* of the image and the *fleshed out content*. The bare bones content of the image consists of the two-dimensional arrangement of geometric color patterns, abstracted from any reference to represented objects. Thus, Kulvicki's bare bones content can be understood in relation to what we previously referred to as the *design properties* (although his stance introduces some distinctions, which I will not take into account, as they are not relevant to the point I want to emphasize). The fleshed out content is the three-dimensional scene compatible with the two-dimensional arrangement of the image's bare bones—which we previously referred to as *pictorial content*. His central claim is that "bare bones contents are context-invariant meanings, or pictorial characters, which in context determine pictorial contents, or pictures' context-sensitive meanings" (2020: 9). The key point lies in the parallel between the distinction of *bare bones* and *fleshed out content* and Kaplan's distinction between *character* and *content*. Just as the character of an indexical, in a given context, determines a content, the bare bones of the picture, in a given context, determines a fleshed out content. Thus, the bare bones content is *context-invariant meaning*, while the fleshed out content is *context-variant* (2020: 9). The bare bones can determine different fleshed out contents, just as an indexical determines different contents in different contexts.

We can illustrate these ideas with an example. The same red ellipse (a design/bare bones content) could be part of an image depicting the top of a red *circular* table, seen from an *oblique angle* (with the red circular tabletop being part of the pictorial/fleshed out content). But this same red ellipse could be part of an image of a red *elliptical* tabletop, seen from above at a *ninety-degree angle*. The geometric arrangement of colors in the image remains the same, but it may represent different things in different contexts.

To understand how Kulvicki relates images to propositions, it is crucial to recognize that, for him, the fleshed out content of a picture is purely *descriptive*—it specifies properties and relations among types of objects without referring to any particular individual. Thus, the fleshed out content of an image would not, by itself, constitute a proposition, but merely a visual description. What is missing is the mapping of this content onto the world. The way this mapping is carried out may treat the descriptive content as either a *definite* or an *indefinite* description, following Bertrand Russell's classic distinction in "On Denoting" (1905).

An indefinite description refers to some object that satisfies a given set of conditions, without identifying which particular one (for example, "a man wearing a hat"), whereas a definite description refers to a unique individual that meets those conditions ("the man wearing a hat").[16] Applied to pictures, this means that the same image can serve either function: it may depict a generic configuration that fits many possible referents (indefinite), or it may be used, in context, to single out a specific referent (definite), depending on how the image is employed and interpreted within a communicative act. But, according to him, *the descriptive fleshed out content of the image is neither definite nor indefinite*.

Thus, for Kulvicki, pictures, by themselves, do not express propositions. *It is the use of descriptive content that determines a proposition*. Once the image's bare bones are mapped onto fleshed out content, they determine a descriptive content, and the use of this pictorial content, "*embedded in a larger communicative act* [...] can be part of the expression of a *proposition*" (Kulvicki 2020: 48—emphasis added). Thus, it is the contextualized use that determines whether the descriptive content of the picture would be regarded as true if satisfied by "some" (as an indefinite description) or satisfied by "one and only one" (as a definite description).

We can translate Kulvicki's ideas into the terms developed in the previous chapters as the thesis that the mapping from design onto pictorial content determines

[16] In "On Denoting" (1905), Russell distinguishes between indefinite and definite descriptions. An indefinite description like "a man wearing a hat" is analyzed as an existential quantification, asserting that there exists at least one individual who satisfies the condition: $\exists x\ (Mx \wedge Hx)$. In contrast, a definite description such as "the man wearing a hat" asserts the existence and uniqueness of an individual who satisfies the condition: $\exists x\ [(Mx \wedge Hx) \wedge \forall y\ ((My \wedge Hy) \rightarrow y = x)]$.

only descriptive content—depicting objects and properties arranged in a certain way. It is the contextualized use that maps this pictorial content onto the domain of referents, thereby determining the denotative content of the image and, through this, making the use of the image the determination of propositional content.

Although Kulvicki formulates his theory as part of pictorial semantics, the centrality of the idea of *use* and *context of use* allows us to frame his position (even if contrary to the letter of the text) in terms of *near-side pictorial pragmatics* (still doing justice to the spirit of the text—so I hope). From this pragmatic perspective adopted here, the question of the relationship between pictorial content and propositions should not be formulated as the metaphysical problem of whether pictures, in themselves, are propositions, but whether we *use* pictures in interlocutory contexts to express propositions. It is the use of a picture embedded in a larger communicative act that determines a proposition as the foundational layer of information in the communicative act. This *pragmatic turn* will enable us to adopt a perspective that shifts the order of priorities and provides a pathway around some central debates in the philosophy of photography.

The *pragmatic turn* allows us to avoid the metaphysical problem of whether photographs are representations that can be true or false. A common viewpoint, epitomized by the Orthodox *motto* "photos do not lie," suggests that photographs are inherently fictionally incompetent. However, from the communicational standpoint, it is unnecessary to wrestle with the issue of whether photographs are intrinsically capable of truth and falsehood, independent of any context of use. This line of inquiry parallels the question of whether a sentence, in the absence of any usage, expresses a proposition that is capable of being true or false. Instead, the more pertinent question becomes: are there instances where a user of a photograph employs it to communicate to a viewer a propositional content, committing to the truth of the conveyed pictorial content (or a part thereof), thereby enabling us to classify the content as true or false? The answer is certainly affirmative. Numerous instances of photographic usage aim to convey to the viewer a specific propositional content that is amenable to assessments of truth or falsehood. (We will later classify these uses as *assertive illocutionary pictorial acts*.)

However, the idea that we use pictures to express propositions should not be interpreted as the thesis that the pictorial content can be *completely* expressed through propositions. Photographs convey a wealth of visual information, making it challenging, perhaps even impossible, to fully articulate their content propositionally. Take, for instance, a high-resolution aerial photograph of a city. The complexity of such an image is staggering.

Kulvicki is cautious on this point. According to him:

> A typical picture might outrun the lexical resources of any given language. Even a lexically mighty language might fail to capture pictorial character or content fully because they are too rich for anything but indefinitely long descriptions. (2020: 33)

The problem may extend beyond mere difficulty to a fundamental lack of conceptual tools for completely naming the properties and describing the objects that make up the pictorial content. This is particularly evident in the representation of colors, shapes, and visual properties—aspects of the represented objects for which language often lacks concepts. Gareth Evans notes that "the content of even the simplest photograph will be [...] something that can be specified *conceptually* only with some loss" (1982: 125).

Setting aside the problem of fully expressing the pictorial content of an image through propositions, the complexity of the fraction that is expressible is still staggering. According to Grzankowski, "It might be helpful to think of a picture as similar to book" (2015: 160). This perspective aligns with the popular maxim that a picture is worth a thousand words. Under this analogy, an image would have content similar to an extensive conjunction of propositions, each expressing aspects of the pictorial content.

But given this complexity, when we point to a photograph and say "this is possible," which of the numerous propositions that are part of the pictorial content is being asserted as possible? Is it all of them, some, or just one? And when we say, pointing to a photograph, "this is not what happened," are we negating the totality of the propositions that make up the pictorial content, or only a select few, or perhaps just one?

The answer to these queries largely hinges on the *context* in which the picture is used and the *communicative intention* of the user. When using a picture in a dialogic context and pointing to it while stating, "this is not what happened," there must be an implicit understanding that both the user and the viewer are in agreement about the subject and are aware of the properties relevant to the context. Hence, there is an underlying assumption that both parties recognize which among the many possible propositions determined by the use of the pictorial content is being addressed in that specific context. This illustrates how the semantics of pictures is intrinsically linked to what we might call *near-side pragmatics*. Here, *informative intention* becomes paramount to communication, even in the context of basic pictures.

The key observation here is that by examining the interlocutory contexts of photographic use, we can avoid the need to answer the metaphysical question of whether the pictorial content is propositional or not. In interlocutory contexts, we *use* pictures to convey propositional contents, and that suffices for a theory of communication. This pragmatic perspective also allows us to overcome the problem of the propositional complexity of pictorial content and the absence of syntactic elements that determine what is being referred to and which properties are being attributed. When used in an interlocutory context, there must be an agreement between the user and the viewer regarding which of the numerous propositions expressible through the pictorial content the picture is intended to convey.

It is also the contextualized use that, according to Kulvicki, allows the specification of whether the descriptive content of the image is used as a definite or indefinite description, even if we lack pictorial elements to syntactically mark this distinction. According to Kulvicki, the determination can be made through the use of images alongside linguistic acts or as something presupposed in the context of use. In the example of the photographic sandwich menu (analyzed in Sect. 2.3), it is presupposed in the context of use that the image is an *indefinite description*—that can be satisfied by various objects.

But the uses of images as definite descriptions (to be satisfied by *one and only one*) or indefinite descriptions (to be satisfied by *at least one*) are attributive uses in which the reference is determined satisfactionally. A picture in these uses would not have a singular content.[17] Kulvicki explains the transition from attributive to referential use, in which images would have a singular content, by drawing on the distinction made by Keith Donnellan (1966) and Kaplan (1978). As seen in Sect. 2.2, in referential use, the pictorial content of the image is *rigidified* when used to refer to a certain object, even if the object does not satisfy the descriptive content of the image. One way this rigidification occurs is through intention as a mechanism of reference. There would be an intention to use the image to select a certain object as part of the propositional content, thus giving the proposition a singular content.

Kulvicki conceives this denotative mechanism in relation to pictures as the repetition of the relationship between *character* and *content*, but on a *second level*, which takes the descriptive fleshed out content as character and determines an individual object as content. On the first level, the bare bones content of the image functions as the character which, in different contexts, can be mapped onto different fleshed out contents. On the second level, a particular descriptive fleshed out content of the image is taken as the character, which, in a given context, can be mapped onto a singular object.

The inherent complexity of images allows part of the image to be used referentially, denoting a singular object, while other parts of the image function as properties attributed to the denoted singular object. Thus, the image could serve the dual function of selecting a singular object and attributing properties to it (in a manner similar to what Novitz suggests (1977)—as we will see in Sect. 3.2.1).

Another second-level operation that takes the pictorial content as character and, in a context, can determine singular content is *iconography*. In the iconographic use of an image, a certain pictorial content is correlated, by rules and conventions, to a singular object. In iconographic reference, denotation does not occur through the specification of a visual description of the object (as in the case of pictorial content taken as an indefinite or a definite description). In the iconographic case, the pictorial content of the image represents elements that function as conventional symbols denoting a specific entity. The descriptive elements of the picture's fleshed out content are used as conventional symbols denoting a singular object.

The third way to understand the singular content of images is based on Kulvicki's analysis of maps and how this analysis applies to photographs. Maps do not denote whatever satisfies their descriptive content. Maps are about specific places, even if the map is imprecise or contains errors. Maps denote by directly referring to a set of locations. The parts of the map denote locations, and the other elements of the map attribute properties to the determined locations. Kulvicki applies this reasoning to photographs, treating the parts that constitute the photographic image as directly referring to parts of the locations denoted by the picture (2020: 121). Thus,

[17] For the distinction between definite descriptions and singular propositions as applied to pictures, see Sect. 2.2.

"locations [in the photograph] represent locations, and their other resources tell us what these locations are like" (2020: 121).

This reasoning would be present in the idea that parts of the photographic image causally depend on parts of the depicted scene. In this case, causality could be used to determine which singular objects are denoted by the picture. Causality would thus play a *metasemantic* role (prior to the semantic content of the picture), since the determination of the referent would occur before the determination of the propositional content of the image. But Kulvicki does not resort to singularization via causality, as his theory focuses on what happens at a *semantic level* (not on what occurs prior to determining the semantic content). What matters for his purposes is that (regardless of the metasemantic mechanisms of fixing the referent—causal or not) the parts of the photograph function as direct referential terms, referring to locations of the depicted scene. Thus, the parts of the bare bones content of the image refer to parts of the space denoted by the image. The descriptive dimension of the image can attribute properties to the objects referred to by the parts of the image.

It is not my intention to analyze all the details of Kulvicki's theory or to adopt all elements of his formulation. My aim is to highlight that the mechanisms of transitioning from bare bones to fleshed out content (or from design to pictorial content), and the second-level mechanisms, which take the fleshed out content as character to determine singular content (by rigidification, iconography, or causality as a metasemantic mechanism), are *context-sensitive*. Thus, when someone in a particular context uses a photograph ostensively and intentionally to communicate something to a viewer, there must be agreement between them for the use of that photograph to determine the same (or sufficiently similar) propositional content. This is precisely what enables photographs (and pictures more broadly) to function as the foundational layer of information in a communicational act.

References

Danto, A. C. 1981. *The transfiguration of the commonplace: A philosophy of art*. Cambridge, MA: Harvard University Press.

Danto, A. C. 1992. Animals as art historians: Reflections on the innocent eye. In *Beyond the Brillo box: The visual arts in post-historical perspective*, 14–31. New York: Farrar, Straus, Giroux.

Donnellan, K. 1966. Reference and definite descriptions. *Philosophical Review* 75 (3): 281–304.

Evans, G. 1982. In *The varieties of reference*, ed. J. McDowell. Oxford: Clarendon Press.

Fodor, J. 2012. Déjà vu: All over again: How Danto's aesthetics recapitulates the philosophy of mind. In *Danto and his critics*, ed. Rollins, 2nd ed. Oxford: Blackwell Publishing.

Gombrich, E. 1973. Illusion and art. In *Illusion in nature and art*, ed. R. L. Gregory and E. H. Gombrich, 193–243. London: Duckworth.

Gombrich, E. 1977. *Art and illusion: A study in the psychology of pictorial representation*. 5th ed. London: Phaidon.

Goodman, N. 1976. *Languages of art: An approach to a theory of symbols*. 2nd ed. Indianapolis: Hackett.

Grice, P. 1957. Meaning. *The Philosophical Review* 66 (3): 377–388.

Grzankowski, A. 2015. Pictures have propositional content. *Review of Philosophy and Psychology* 6 (1): 151–163.
Hatfield, G. C. 2009. *Perception and cognition: Essays in the philosophy of psychology*. Oxford: Oxford University Press.
Hopkins, R. 2015. The real challenge to photography (as communicative representational art). *Journal of the American Philosophical Association* 1 (2): 329–348.
Heal, J. 1997. Indexical predicates and their uses. *Mind* 106:619–640.
Kaplan, D. 1978. Dthat. In *Syntax and semantics*, 221–243. Academic Press.
Kaplan, D. 1989. The logic of demonstratives. In *Themes from Kaplan*, ed. J. Almog, J. Perry, and H. Wettstein. Oxford: Oxford University Press.
Korta, K., and J. Perry. 2011. *Critical pragmatics: An inquiry into reference and communication*. Cambridge: Cambridge University Press.
Korta, K., and J. Perry. 2020. Pragmatics. In *The Stanford encyclopedia of philosophy (Spring 2020 Edition)*, ed. E. N. Zalta. Available at: https://plato.stanford.edu/archives/spr2020/entries/pragmatics/.
Kripke, S. 1979. Speaker's reference and semantic reference. In *Contemporary perspectives in the philosophy of language*, ed. P. A. French, T. E. Uehling Jr., and H. K. Wettstein. Minneapolis: University of Minnesota Press.
Kulvicki, J. 2006. *On images: Their structure and content*. Oxford: Clarendon Press.
Kulvicki, J. 2014. *Images*. New York: Routledge.
Kulvicki, J. 2020. *Modeling the meanings of pictures: Depiction and the philosophy of language*. Oxford: Oxford University Press.
Lopes, D. 2004. *Understanding pictures*. Oxford: Oxford University Press.
Lopes, D. 2005. *Sight and sensibility: Evaluating pictures*. Oxford: Clarendon Press.
Lopes, D. 2016. *Four arts of photography*. Oxford: Wiley-Blackwell.
Marr, D. 1982. *Vision: A computational investigation into the human representation and processing of visual information*. Cambridge, MA: MIT Press.
Matthen, M. 2005. *Seeing, doing, and knowing: A philosophical theory of sense perception*. Oxford: Clarendon Press.
Peacocke, C. 1987. Depiction. *Philosophical Review* 96:383–410.
Meskin, A., and J. Cohen. 2008. Photographs as evidence. In *Photography and philosophy: Essays on the pencil of nature*, ed. S. Walden, 70–90. Malden, MA: Blackwell Publishing.
Novitz, D. 1977. *Pictures and their use in communication: A philosophical essay*. Springer.
Pylyshyn, Z. 2003. *Seeing and visualizing: It's not what you think*. Cambridge, MA: MIT Press.
Russell, B. 1905. On denoting. *Mind* 14 (56): 479–493.
Sperber, D., and D. Wilson. 1995. *Relevance communication and cognition*. Oxford: Blackwell Publishers Ltd.
Strawson, P. F. 2003. *Individuals*. London: Routledge.
Zeimbekis, J. 2010. Pictures and singular thought. *The Journal of Aesthetics and Art Criticism* 68 (1): Winter.
Zeimbekis, J. 2015. Why digital pictures are not notational representations. *The Journal of Aesthetics and Art Criticism* 73 (4): 449–453.
Wollheim, R. 1980. *Art and its objects*. 2nd ed. Cambridge: Cambridge University Press.
Wollheim, R. 1987. *Painting as an art*. London: Thames & Hudson.

Chapter 3
Far-Side Pictorial Pragmatics Applied to Photography

Abstract This chapter is dedicated to developing the foundations of a far-side pictorial pragmatics as applied to photography, with the aim of understanding how speech acts can be performed through the use of photographs. To this end, we begin by examining some of the core ideas in John Austin's speech act theory and how they have been extended to the domain of pictures by David Novitz and Rae Langton. Novitz explores how pictures can be used to perform assertive speech acts. Langton, in turn, analyzes pornography in connection with speech act theory and demonstrates how photographs can be used to subordinate and silence women. Her perspective broadens our understanding of the ways in which photographs may function as instruments for authoritative speech acts. I will then argue that photographs can be related to speech acts in distinct ways. The chapter also offers a detailed analysis of John Searle's taxonomy of speech acts, as the basis for a classification of photographic speech acts into five categories: assertive, directive, commissive, expressive, and declaratory. Particular emphasis is placed on the idea that the direction of fit between photograph and world varies depending on the type of speech act being performed. Finally, we turn to Paul Grice's theory of conversational maxims to explore how these principles can be adapted to the domain of photographic communication, thereby allowing us to develop the notion of photographic implicatures.

Keywords Illocutionary photographic acts · Photographic implicatures · Social practices · Direction of fit

The path taken in the previous chapters has allowed us to understand communication as having two layers and to draw a parallel between these layers and two fields of study in pragmatics. In Chap. 1, drawing on the critique of Paul Grice's theory of communication developed by Sperber and Wilson (1995, 2004), we came to understand communication as consisting of *a foundational layer of information* about a given subject, and *a second layer that indicates that this information was intentionally manifested*. Thus, in communication, there are two types of intentions: the

G. Ghisoni da Silva, *The Philosophical Foundations of Photography as a Means of Communication*, Synthese Library 526,
https://doi.org/10.1007/978-3-032-02649-1_3

informative intention (the intention to inform an interlocutor of something) and the *communicative intention* (related to the ostensive use indicating that the first layer was intentionally manifested). In the case of photographic communication, the photograph could play the role of the foundational layer of information, to which a second layer indicating that the first layer was intentionally manifested could be added.

In Chap. 2, we drew a parallel between the two fundamental layers of communication and the two different scopes that pragmatic theories can have, as outlined by Korta and Perry (2011, 2020), exploring how these distinctions can be extended to the pictorial domain. The first, called *near-side pictorial pragmatics*, is dedicated to understanding the contextual elements that determine what the image depicts. The fundamental point is that for photographic images to be used as the foundational layer of information in a communicative act, it is necessary for the user and the viewer to sufficiently share the context-sensitive elements recruited in determining what the image depicts and denotes. In general terms, we can formulate this requirement as the idea that the user and the viewer must agree, to some extent, on the propositional content determined by the contextualized use of the pictorial content.

In Chap. 3, our attention will be devoted to the second layer that constitutes communication. It is at this second level that we will find the distinction between what the image *depicts* (determined within the scope of near-side pictorial pragmatics) and what the use of the image *communicates*. The main problem to be addressed is understanding how to move from depiction to the intentional ostensive use of photographs as a communicative act.

In our previous analyses, the understanding of what happens beyond depiction was initially outlined through the parallel between the two layers of communication and speech act theory. The photograph, as the foundational layer, can play the logical role of the *locutionary act*, and its contextualized use can perform a speech act. What is communicated in a particular use of the photograph is related to the *illocutionary act* performed. We could also find in photographic communication the idea of an *utterance act* and a *perlocutionary act*. Thus, the structure of photographic communication could be thought of as isomorphic to speech act theory:

1. *Photographic Utterance Act*: The user intentionally presents a specific photograph as a candidate for attention to a spectator in a given context.
2. *Photographic Locutionary Act*: (In performing the utterance act), the user depicts a scene in which objects are spatially arranged and endowed with visual and expressive properties—determined in a context-sensitive way (within the scope of near-side pictorial pragmatics).
3. *Photographic Illocutionary Act*: (In performing the locutionary act) the use of the image carries out an action in that context (within the scope of far-side pictorial pragmatics).
4. *Photographic Perlocutionary Act*: (In performing the illocutionary act) the user affects the spectator in a particular way.

Near-side pictorial pragmatics applied to photography aims to understand how the *utterance act* relates in a context-sensitive way to the *locutionary act*. Far-side pictorial pragmatics applied to photography aims to understand how the *locutionary act* relates to the *illocutionary act* and the *perlocutionary* effects.

The understanding of how photographic communication can be thought through these elements will be developed in this Chap. 3, drawing on the theoretical frameworks of John Austin's (1962) and John Searle's (1969, 1979) speech act theories. Subsequently, we will also investigate the parallels between the use of photographs and the maxims and principles of communication proposed by Paul Grice (1957, 1989)—taking into account the critique of Grice discussed in Sect. 1.3. This paves the way for exploring the idea of photographic implicatures and the parallel between implicatures and indirect speech acts.

There is an important theoretical justification for thinking about communication from a dual perspective based on *speech act theory* and *Paul Grice's theory of communication*. As mentioned in previous chapters, the theory of photographic communication proposed in this book will draw a distinction between *conventional* and *non-conventional* contexts. Although actual contexts form a *continuum* with varying degrees of conventionality, these two extremes will allow us to think about how the structure of communication can be conceived in relation to the role of intentionality. I will propose in detail in Chap. 4 that conventional contexts will be those in which participants in social photographic practices share a collective intentionality related to the perceived function of the practice. This perceived function, in turn, determines the purpose of photographs within the practice. We can think of this purpose as the primary speech act (or acts) that the use of photographs aims to perform in that context. Thus, the purpose attributed to photographs within the practice can be understood as the *illocutionary purpose of photographs* in that context. Intending this illocutionary purpose will be a necessary condition for participation in the practice. Therefore, the mere presence of the photograph within a certain practice will be sufficient for the attribution of a communicative intention in line with the illocutionary purpose and function of the practice. In other words, we can adopt a *non-intentionalist* perspective of communication in this case, as the ostensive manifestation of communicative intention by the user is not necessary for communication, since such communicative intention is a presupposed requirement for participation in the practice.[1] We can likewise attribute to all pictures within a photographic social practice a communicative intention aligned with the illocutionary purpose of the practice. We can thus attribute to the image itself, in a derivative manner, the performance of the speech act—allowing us to say that a certain photograph performs that speech act.

Non-conventional contexts will be those in which there is no prior determination of a function, and thus no prior determination of the illocutionary purpose of photographs. For communication to occur in non-conventional contexts, the user of the image must ostensively manifest their communicative intention, and the spectator must apprehend this intention.

[1] This does not mean denying that communicative intention is fundamental, but rather that the ostensive manifestation of communicative intention is not required within a conventional practice, since this intention is already presupposed in the use of photographs within the social practice.

The significance of a dual theoretical perspective lies in the fact that understanding conventional contexts of photographic communication can be achieved through the photographic version of speech act theory, while non-conventional contexts will require an understanding of the role of reflexive intention and how communication is based on principles and maxims, aligned with the photographic version of Paul Grice's theory of communication. This dual theoretical perspective also allows us to explain how it is possible for a photograph to be used with a certain communicative intention, but fail to achieve the intended act or perform a different act than intended. Such an understanding would only be possible if we approach communication from a theoretical perspective that integrates both intentionalist and non-intentionalist approaches to communication.

However, to adequately speak of a *communicative intention*, we need first to understand how it is distinct from the role that intentionality can have in determining pictorial content (as in Richard Wollheim's (1980) theory (Sect. 2.1)) and from the intentionalist approach to pictorial reference (Sect. 2.2). Communication, depiction, and denotation are three distinct levels and can, to some extent, be logically independent.

Pictorial intention can serve as the *standard of correctness* in determining pictorial content, given the underdetermination of the relationship between design and pictorial content (Sect. 2.1). Additionally, pictorial content can be mapped onto existing objects, and one of the mechanisms of pictorial reference is intention, which can *rigidify* the pictorial content (Sects. 2.2 and 2.3). This marks the distinction between the role of intention in the domains of *picture theories* and *theories of pictorial reference*. However, there is yet a third level related to the speech act performed through the use of the picture. Intention can play a crucial role in communication by determining what speech act is intended through the use of a certain picture, with specific pictorial content and denotation. We can use the same picture for numerous illocutionary purposes. But adopting an intentional approach to speech acts does not require or presuppose that we adopt intentional theories of pictorial content or pictorial reference. We could, for instance, in a certain context of use, accept that a photographic image depicts something by triggering the spectator's recognition capacities and that the denotation is fixed using causality as the mechanism. In addition to these positions regarding depiction and denotation, we could assume that the use of this image with a communicative purpose could occur in a way that communication depends on the user's communicative intention.

To develop a far-side pictorial pragmatics applied to photography, it is necessary to understand some central theses of John Austin's (1962) speech act theory and some important developments by John Searle (1969, 1979). The idea of a theory of pictorial communication based on speech act theory is not without its precedents. I will take David Novitz (1977) and Rae Langton (2009) as examples of pictorial pragmatic theories, which provide us with a starting point for applying Austin's theory to pictures (Sect. 3.2).[2] This will give us a path to constructing a far-side pictorial pragmatics applied to photography. The developments I will propose based on Novitz and Langton's insights will be achieved by expanding the theoretical horizon beyond Austin's theory, incorporating the contributions of John Searle. It is from Searle's taxonomy of illocutionary acts and his understanding of the dimensions of variation of speech acts that I will build a taxonomy of photographic illocutionary acts and the foundations of an ontology of photographic social practices (to be explored in Chap. 4). Another important point in our analysis in this chapter will be to understand how photographs and speech acts can be related in different ways, in order to identify which of these ways are relevant and central to the purposes of this book. I will conclude Chap. 3 by developing the photographic version of some central ideas of Paul Grice's theory.

3.1 John Austin's Doctrine of Infelicities

John Austin's (1962) starting point is the distinction between two uses of language. In *constative* utterances, the words express propositions that may or may not describe the world correctly. In *performative* utterances, the utterance indicates that the speaker is performing an act. Performative utterances begin with expressions like "I declare," "I promise," "I bet," "I swear," "I order," "I predict"—always in the first person, the present tense, the indicative mood, and the active voice. Unlike constative utterances, which can be true or false, performative uses can succeed or fail in performing the intended action. For example, someone who is not a judge cannot successfully perform the speech act of declaring a defendant guilty by saying "I declare the defendant guilty."

Although the distinction between constative and performative uses is Austin's point of departure, performative utterances do not play a fundamental role in his theory since speech acts can be performed without using a performative utterance. One can promise something without saying "I promise," place a bet without saying "I bet," or take an oath without stating "I swear." In some cases, no utterance is necessary at all, as he considers the possibility of speech acts being performed without verbal utterances, such as making a bid at an auction by raising one's hand. However, his study primarily focuses on verbal utterances, leaving it to the reader to

[2] It is not my intention to deny that there are other contributions equally significant, or even more so, to the parallel between pictures and speech acts. Søren Kjørup (1978) and Daisy Dixon (2018) are examples of highly innovative approaches with many contributions that could also have been included in the analyses presented in this book. However, my goal is not to provide a complete survey of the history of the relationship between pictures and speech acts. I will use Novitz and Langton as examples because I believe their proposals offer a comprehensive perspective on how to connect and deepen this relationship, facilitating the further developments I will propose through the analysis of John Searle's ideas.

understand how these ideas might apply to non-verbal cases (as in the parallel that will be drawn here with the use of photographs).

The fundamental point of the performative uses of language is that, in these cases, *in* saying something, an *action* is performed. The act of saying something Austin calls a "locutionary act," and the act performed *in* saying is the "illocutionary act" (1962: 94–107). The performance of an illocutionary act can have effects, which the author calls "perlocutions". He provides the following example: by performing the *locutionary act* of saying "Shoot her!," the speaker performs the *illocutionary acts* of advising and/or ordering the interlocutor, which may have the *perlocutionary effect* of persuading him to shoot her (1962: 94–107). The expression "speech act" is used synonymously with the expression "illocutionary act."

The author describes the illocutionary act as the "force of the utterance" (1962: 72). Different illocutionary forces can be added to the same locutionary act—for instance, the locution "close the door" can be used as a request, an order, a plea, a challenge, etc., where different *illocutionary forces* are attributed to the same locutionary act.

The study of speech acts extends beyond the study of performative utterances and aims to understand the conditions for the successful *uptake* of illocutionary acts. There are requirements that must be met for an illocutionary act to take effect from the utterance of a locutionary act. The *doctrine of infelicities*, proposed by Austin, aims to understand the different ways in which a speech act can fail.

The first two requirements imposed by Austin for the successful performance of illocutionary acts are as follows:

> (A.1) There must exist an accepted conventional procedure having a certain conventional effect, that procedure to include the uttering of certain words by certain persons in certain circumstances, and further,
>
> (A.2) the particular persons and circumstances in a given case must be appropriate for the invocation of the particular procedure invoked. (1962: 14–15).

In more general terms, there must be a practice of language use in which a certain convention is accepted, whereby a certain utterance, made by the appropriate people, under the correct circumstances, and according to specific procedures, has certain effects. For example, for the performative "I declare the defendant guilty" to trigger the perlocutionary effect of the defendant's imprisonment, the declaration must not only be made by the appropriate person but must also occur within an institution, under certain circumstances, and following procedures that determine that the locutionary act enacts the verdict of condemning the defendant. Failure to meet these felicity conditions results in the failure of the speech act, as it does not achieve *uptake*. There are different ways in which a speech act can be *infelicitous*.[3]

Two additional requirements are imposed by the author:

[3] One point to note in this initial set of requirements is the necessity of a procedure *conventionally accepted* by language users. This element will be relevant for our subsequent analyses of the use of photographic images non-conventional contexts, for which an approach distinct from Austin's will be necessary.

(B.1) The procedure must be executed by all participants both correctly and

(B.2) completely. (1962: 15).

For a bet to be made, for example, it is not enough for the speaker to say, "I bet with you that Brazil will win the next Olympic gold medal in floor gymnastics," but the interlocutor must accept the bet for the speaker's illocutionary act to be implemented. In cases like this, the uptake of the speech act does not depend solely on the speaker, the interlocutor must also fully execute the procedure—accepting the bet proposed by the speaker.

Failures related to requirements of type (A) and (B) would render the speech act not successfully performed at all, such that *uptake* of the act would not occur. The act may be intended, words may be uttered, but without fulfilling the requirements, the act is not established and becomes *null* or *void*. For example, a trial that does not follow the required procedures would be null, and the pronouncement of a sentence without the correct procedures would not have the *force* to establish the verdict of condemning the defendant or set in motion the perlocutionary effect of their imprisonment. I can throw a bottle against the hull of a ship and say, "I christen this ship 'Antônio Carlos Jobim,'" but if I do not have the necessary authority, my words do not have the *force* to christen the vessel.

An additional type of requirement imposed by Austin includes psychological requirements, which may extend to future moments after the locutionary act. According to Austin:

(Γ. 1) Where, as often, the procedure is designed for use by persons having certain thoughts or feelings, or for the inauguration of certain consequential conduct on the part of any participant, then a person participating in and so invoking the procedure must in fact have those thoughts or feelings, and the participants must intend so to conduct themselves, and further

(Γ. 2) must actually so conduct themselves subsequently. (1962: 15).

For example, for the performative "I promise that I will help you tomorrow" to be an actual promise, it must be made *sincerely* by the person who utters it. Sincerity is a psychological requirement for commissive acts (where the speaker commits to an obligation towards the interlocutor) and for expressive acts (where the speaker apologizes, thanks, offer condolences, etc.).

An important difference between requirements of type (A), (B), and type (Γ)—expressed by Austin in the choice of the Greek letter—is that the failure to meet the first set of requirements prevents the implementation of the speech act. In the case of failing to meet requirements of type (Γ), the speech act is implemented (and the uptake can occur) but becomes *empty* (rather than *null*). An insincere apology is a speech act that may be accepted by the interlocutor as an apology (thus implementing the act), but it becomes empty due to insincerity.

Requirement (Γ.2) concerns cases where there is a need to maintain the psychological requirement for future moments following the speech act. A performative "I promise that I will help you tomorrow," made sincerely, but where the speaker later revises their intentions to fulfill the promise (so that they no longer intend to keep it), renders the previous promise empty.

Austin outlines a typology of infelicities. He distinguishes infelicities of types (A) and (B), which he calls "Misfires" (cases where the acts are not implemented), from infelicities of type (Γ), which he calls "Abuses" (cases where the act is implemented but becomes empty). Infelicities of type (A) are classified as "Misinvocations" because they invoke an act without the power to implement it. Austin calls this "Misapplication" cases where utterances are performed by inappropriate people or procedures (A. 2). Failures of type (B), termed "Misexecutions," are divided between "Flaws" (when the act is not performed by all participants (B. 1)) and "Hitches" (when performed incorrectly or incompletely (B. 2)). The failure to meet psychological requirements (present in (Γ. 1)) is described as "Insincerities."

The names assigned to the types of infelicities are not crucial to our study since the typology is incomplete—Austin did not find names for the infelicities of type (A. 1) and (Γ. 2). What is important to note is that different types of infelicities are related to different requirements for the implementation or maintenance of the implemented act. And different types of illocutionary acts will have different requirements.

Austin proposes an initial classification of illocutionary acts, being aware of the provisional and insufficient nature of his classification. Speech acts can be classified as follows:

1. Verdictives: delivering a judgment, official or unofficial—condemning, judging, assessing, evaluating, etc.
2. Exercitives: exercising power, rights, or influence—appointing, voting, ordering, urging, advising, warning, etc.
3. Commissives: committing to an obligation or declaring intention—promising, swearing, betting, planning, etc.
4. Behabitives: engaging in social behavior—apologizing, thanking, commanding, welcoming, congratulating, praising, etc.
5. Expositives: performing an act in the course of an argument or conversation—declaring, assuming, postulating, contesting, denying, etc.

From these ideas, Austin proposes two fundamental theses that provide the basis for the theory of speech acts[4]:

(a) The focus of philosophical analysis of language is not the proposition (which can be true or false), but the *total speech act*. "The total speech act in the total speech situation is the only actual phenomenon which, in the last resort, we are engaged in elucidating" (1962: 147).
(b) "Stating, describing, &c., are *just* two names among a very great many others for illocutionary acts; they have no unique position". (1962: 147–48)

Subsequently, our task will be to explore how the distinction between locutionary, illocutionary, and perlocutionary acts applies to the use of photographs, enabling us

[4] Austin presents various theses in his analysis of Chap. XII. I will address only those that are relevant to understanding communication through the use of photographs.

to comprehend the distinction between what the image depicts and what it communicates. It will also be crucial to understand how photographic speech acts may have requirements that, if unmet, could lead to the non-implementation of the speech act or render the act empty. This understanding will allow us to delineate the different conditions for the uptake of photographic acts as well as a doctrine of infelicities.

3.2 Applying Austin to Pictures

In this chapter, we will focus on analyzing two proposals for applying John Austin's theory to pictures. Several authors have addressed this topic (such as the contributions of Søren Kjørup (1978) and Daisy Dixon (2018)). However, priority will be given to the formulations of David Novitz (1977) and Rae Langton (2009), as they explore themes in depth that will be highly relevant for the subsequent analyses in this book.

3.2.1 Pictorial Illocutionary Acts in David Novitz

In a highly innovative way, one of the first in-depth explorations of the parallel between the use of pictures and John Austin's speech act theory was conducted by David Novitz (1977), in his theory of *pictorial communication*. Although this theory is quite limited, as it focuses primarily on *assertive* pictorial speech acts, it includes several significant contributions that deserve attention. According to the perspective proposed in this book, we could attribute to Novitz the merit of having developed one of the first *far-side pictorial pragmatic theories*.

According to Novitz, the same picture can be used for different purposes:

> A picture, say of a man, can be used in an indefinite number of ways. It can be used to conjure up memories, to advertise the clothes that he wears, to inform an audience of a rare disease, or to give a visual description of the symptoms of infective hepatitis. It can be used as an object of worship, as a decoration; or again, it can be used to explain an event in history or even to identify a criminal. (1977: 67)

Given this plurality of communicative uses of the same picture, it becomes necessary to distinguish between what the picture *depicts* and its *use*. It is also necessary to distinguish between the *picture's production* and its *use*, as many pictures are used for purposes different from those for which they were produced.

Even though a picture can have different uses, for it to be used *as a picture*, its pictorial content must have some relation to the intended communicative act. This creates a tension between how the pictorial content limits the uses of the picture, even though different uses are still possible. There are limits to the uses to which we can assign a certain picture. As Novitz emphasizes, "One cannot (…) use a picture of Queen Victoria on horseback in order to illustrate Alice's tea party" (1977: 68).

For such a use to be possible, we would have to convert the picture of the queen into a *symbol* that, in an arbitrary and conventional way, represents Alice's tea party (this could be a case akin to an iconographic reference, as analyzed in Sect. 2.2). In this case, it would not be the *pictorial content* of the image that determines what it represents, but a convention among the speakers. On the other hand, in philosophically interesting cases of picture use, "pictures are used to convey information which pertains to, or is about, whatever is pictured" (1977: 68). In these cases of *pictorial* (non-symbolic or conventional) *representation*, the content of the picture plays a communicative role related to the intended use. However (and this is the point to be noted), even if there is this kind of limitation regarding pictorial representation, the same picture of Queen Victoria can still be used for numerous distinct pictorial purposes related to its pictorial content.

We can approach this tension by noting that the communicative act is, for Novitz, a *necessary* but not *sufficient* condition for pictorial representation. Novitz's idea is that the *use* of a picture is a *necessary condition* for pictorial representation. It is only in relation to a certain use of the picture that it becomes a representation that *describes*, *illustrates*, *informs*, etc. However, although the use of a picture is necessary for it to be a representation, it is not *sufficient*. For pictorial representation to occur (in a non-symbolic or non-conventional way), the use must be related to the content of the picture or aspects of the content.

The connection with Austin's speech act theory is proposed by Novitz in the following terms:

> [T]o represent something pictorially is to perform a pictorial act of communication which pertains to, or is about, what is pictured (or an aspect of what is pictured), where such an act may be performed under a range of descriptions commonly given to *illocutionary acts*. (1977: 69 – emphasis added).

The central thesis put forward by Novitz is that when we use an image for a communicative purpose related to its pictorial content, we perform a speech act that can be characterized, following John Austin, as a *pictorial illocutionary act*.

Novitz's theory of pictorial communication is based on John Austin's distinction between the *uttered sentence* (the locutionary act) and the *illocutionary act* performed (also characterized as illocutionary *force*). Novitz argues that Austin, although prioritizing illocutionary acts in which locutionary acts are sentences, would be open to granting the possibility of non-verbal, perhaps pictorial, illocutionary acts (1977: 77). Building on this possibility, Novitz develops the idea that images could be used with the intention of producing a certain cognitive response in an audience. In such cases, the image would fulfill the role of the *locutionary* act, the use of the image in context, with a certain intention, would be the performed *illocutionary* act, and the effect of using the image would constitute the *perlocutionary* act. In other words, images could be used to inform, order, warn, etc., and these actions could have effects.

Although Novitz acknowledges that not every illocutionary act involves referring to a subject and attributing properties to them (as one can, for example, greet someone by saying "hello"), his analysis is restricted to *propositional* pictorial illocutionary acts. His analysis focuses on cases where pictures are used as part of illocutionary acts to refer to a subject and predicate something about it, such as qualities, properties, relations, actions, or states (1977: 89–90).

One problem with the proposed parallel between illocutionary acts and pictorial illocutionary acts is that pictures do not have a grammar analogous to the surface grammar of propositions in ordinary language (as we analyzed in Sect. 2.5). In propositions, we can determine what is being referred to and the content is being attributed through the subject-predicate form and formal syntactic characteristics. However, there are no pictorial nouns or pronouns, pictorial verbs, or verb inflections that allow for an analogous function to verb tenses in images. Thus, in an image, the syntactically relevant elements of the design do not clearly determine the referred entity and the predicative content attributed to it. Many different propositions can be extracted from the same picture. However, according to Novitz, this does not mean that images cannot express propositions in certain contexts of use. According to him (adopting a perspective similar to the pragmatic perspective proposed in Sect. 2.5): "the context, one might say, provides certain clues as to how the image is being employed: what illocutionary act its use performs and what proposition it is made to express" (1977: 92).

The importance that Novitz attributes to speech acts aims to dispel the misconception that sentences and pictures would express propositions independently of their contextualized uses. Regarding language, the formal characteristics of a sentence are not sufficient to express a proposition: "any sentence which expresses a proposition has to be used in an illocutionary act in order to do so" (1977: 91). Just as it is the act of using a sentence that enables it to express a proposition, for Novitz, it is the act of using an image, in a context, that allows it to express a proposition. Therefore:

> [T]he same picture used in the absence of the relevant contextual features cannot be said to express the same proposition, we may assume that the context in which a pictorial illocutionary act is performed is to some extent a determinant of, or is responsible for, the propositional content of the act. (1977: 92).

In other words, it is the use of a picture to perform an *illocutionary* act that partly determines the locutionary act of the picture as a propositional content. Thus, from Novitz's perspective, the context-sensitive elements that are part of the scope of *near-side pictorial pragmatics*, which concern the contextualized determination of the propositional content of the pictorial content, will only be activated once the picture is used, within the scope of *far-side pictorial pragmatics*, as an illocutionary act. Thus, in addition to the elements that determine the transition from design to pictorial content and from pictorial content to denotation being context-sensitive, for the picture to determine a propositional content, it is necessary to *use* the picture intentionally as part of an *illocutionary act*.

The mistake of assuming that pictures could express propositions independently of illocutionary acts is traced by Novitz to the fact that pictures are often artifacts created with a specific communicative use in mind. A certain picture type is generally used to express a certain type of proposition. For example, pictures in a botany book are generally used to indicate certain plants and attribute certain properties to them. A viewer familiar with this use, when faced with a typical picture from a botany book, would know how to "read" the proposition expressed through the use of that picture. But this does not mean that the image already carries this propositional content independently of any use. There is an assumption by the viewer that the image was produced with the intention of being part of the illocutionary act in which it is used to express such a proposition.

One aspect that justifies the parallel proposed by Novitz between pictures and speech act theory is that the uses of pictures often aim at perlocutionary effects. When using a certain picture in a context, expressing a certain proposition (in which a subject is indicated and properties attributed), there is often an intention to produce a certain cognitive response in the audience. Additionally, there may be the explicit intention that the production of this response is understood by the viewer as something intended (thus fulfilling the requirement imposed by Peter Strawson of an "audience-directed intention" (1964: 459)—which Jerry Fodor calls a "reflexive intention" (2012: 63)).

Novitz distinguishes between the *perlocutionary* effect, the *aesthetics* of the image, and its *expressive* dimension. The three cases concern what pictures can elicit. For Novitz, the perlocutionary act is restricted to the effect caused by the use of the picture as part of the illocutionary act, in which a proposition is asserted through the use and there is an explicit intention to cause a certain cognitive response in the viewer through the use of the proposition. On the other hand, even if a picture is not part of an illocutionary act, it can, for example, still please the viewer due to the arrangement of colors, be delight due to the geometry of the composition, be delicate, elegant, etc. (1977: 139). In the terms proposed in Chap. 2, we can say that the aesthetic experience of the expressive properties of the picture (with its different modes, foci of attribution, and mechanisms of pictorial expression) can be distinguished from the perlocutionary effects of the picture's use. Thus, the perlocutionary effect is restricted to cases where there is the performance of a pictorial illocutionary act. The aesthetic and expressive dimensions are, thus, independent of this use. According to Novitz:

> Expressive properties are ascribed to pictures independently of how they are used: that they are ascribed in virtue both of the visual qualities of pictures and of the historical and stylistic context in which they were produced – but never in virtue of their use in an illocutionary act. (1977: 150)

I will agree with Novitz's general idea that *there can be* an independence between the expressive properties of the picture and the speech act performed, but I will diverge by adopting a different scope regarding the types of speech acts that can be attributed to the use of photographs. Novitz is restricted to illocutionary acts that, according to Searle's taxonomy (Sect. 3.2), we can call *assertive*. In these cases, the illocutionary use of the picture would be distinct and could be independent from the attribution of expressive properties. However, I will argue (in Sect. 3.5.4) that we can also perform expressive illocutionary acts through the use of photographs, such that, in this case, the expressive properties and the performed speech act are not independent and can vary depending on the user's communicative intention (which may require expressiveness be attributed in a specific way). By extending my analysis to cases beyond the restricted scope of Novitz's analysis, I will not agree with his assertion that the expressive properties are ascribed "*never* in virtue of their use in an illocutionary act."

3.2.2 Pornography as a Speech Act in Rae Langton

Although David Novitz's work provides an important and highly original starting point for constructing a theory of photographic communication based on speech act theory, the most significant articulation of the relationship between pictures and speech acts can be found in Rae Langton's work on pornography as a speech act. Additionally, unlike Novitz, who addresses communication considering pictures in general, Langton's work allows for a more direct connection with the case of photography. Her analysis concerns pornography in general, but pornographic photographs play a central role.[5] But the most important point is that Langton moves us beyond the *assertive* illocutionary acts (in which pictures are used to express propositions, as in Novitz's analysis) by arguing that photographs can perform *verdictive* and *exercitive* acts (according to John Austin's taxonomy (Sect. 3.1)).

In her famous paper from 1993 "Speech Acts and Unspeakable Acts", Rae Langton starts with a dual characterization of pornography.[6] For those who claim that pornography should be protected by the First Amendment to the United States Constitution, which guarantees freedom of speech, pornography could be characterized as a form of *speech*. According to Catharine MacKinnon (1987), pornography could also be characterized as a form of *action*, as it *subordinates* and *silences* women. Rae Langton's thesis connects these two characterizations and suggests that we understand pornography as a *speech act*, as it is speech that performs an action.

Applying Austin's speech act theory to the understanding of pornographic photographs, Langton distinguishes what the pictures depict (the *locutionary* act), the action performed by the pictures (the *illocutionary* act), and their consequences (the *perlocutionary* act). From a locutionary perspective, pornographic photographs depict women in a condition of subordination and produce the perlocutionary effect of perpetuating subordination by encouraging the performance of the depicted acts of subordination. Subordination is understood as the act of placing someone "in a position of inferiority or loss of power, or to demean or denigrate them" (Langton 2009: 35). By perpetuating subordination, pornography not only depicts the subordination of women but also *causes* subordination as a perlocutionary effect.

---

[5] Throughout this chapter, I will formulate Langton's theses exclusively in relation to photography, although they can be extended to other forms of pornography.

[6] The citations and references will be given according to the reprint of the article in Langton's 2009 book.

However, besides depicting subordination and having the perlocutionary effect of causing it, the author aims to defend the central thesis that "pornography is, in and of itself, a form of subordination" (2009: 26). In other words, besides the *locutionary* and *perlocutionary* acts, pornography would perform the *illocutionary* act of subordinating women. Langton also aims to defend MacKinnon's additional thesis that pornography *silences* women, showing that the silencing is not metaphorical (2009: 26). Pornography performs the illocutionary act of *silencing* by preventing women's locutionary acts from performing illocutionary acts and having perlocutionary effects. These two theses identify conflicts of distinct natures. In the case of subordination, the supposed right of men to produce and consume pornography conflicts with women's right to the same civil status as men. In the case of silencing, the supposed right of men conflicts with women's freedom of speech (2009: 29–30).

The speech acts of subordination and silencing performed by pornographic photographs could be classified respectively, according to Austin's theory, as *verdictive* and *exercitive* acts. Verdictive acts typically classify, rank, or determine the value of something. In many acts of classification, ranking, and value determination, one of the felicity conditions is that the person performing the act has the necessary *authority*. The speaker must fulfill an authoritative role to have the power to perform these acts. This condition also applies to exercitives, which are acts performed when issuing an order, prohibition, authorization, promulgation of a law, silencing, etc. According to Langton:

> [T]he crucial feature of verdictive and exercitive illocutions is their sensitivity to the speaker's authority, and we can accordingly group them together under the label authoritative illocutions: actions whose felicity conditions require that the speaker occupy a position of authority in a relevant domain. (2008: 37)

According to Langton, what turns a particular picture that depicts women in conditions of sexual subordination into one that performs the illocutionary act of subordinating women cannot be explained solely by its pictorial content. As the author states: "an utterance's depicting subordination is neither necessary nor sufficient for its having the force of subordination" (2009: 26). Photographs that depict women in conditions of sexual subordination can be used in non-pornographic contexts ("documentaries, for example, or police reports, or government studies, or books that protest against sexual violence, or perhaps even legal definitions of pornography" (2009: 38)). It is the contextualized *use* to which the locution is put that makes a particular picture the performance of a certain illocutionary act. This also allows us to note that the point in dispute does not concern merely what is depicted and how it is depicted (the locutionary act), but rather the illocutionary act performed through its use.

The central difficulty is that, in many cases, determining which speech act is performed is not something that can be easily accomplished. In the paradigmatic cases of linguistic speech acts, the felicity conditions of the act are clearly determined in relation to the conventions and rules of the context. The context determines that, by performing a certain locutionary act and fulfilling certain conditions, an illocutionary act is instantiated (the *uptake* occurs) and will have certain perlocutionary effects. However, ordinary contexts of use rarely have clearly defined conventions, and the determination of felicity conditions is always carried out with a certain degree of vagueness and flexibility.

As mentioned, subordination is paradigmatically the act of ranking someone in a position of inferiority. It is an act in which typically the one who ranks must occupy a position of authority. One of the difficulties in determining whether pornography performs the illocutionary act of subordination is the fact that it does not meet all the felicity conditions of paradigmatic cases of subordination. However, Langton argues, it can still be understood as the act of subordination (2009: 43). The author explores three lines of argument to support this claim.

In many cases, we can trace the effects to the illocutionary act performed by conceiving the act as the cause of the effect. From this perspective, what would explain the perlocutionary effect of pornography causing and perpetuating subordination is the fact that it is, in itself, an act of subordination. But the inference of the act from the effect is fallible, and there are cases where this relationship does not hold.

The second line of argumentation starts from the observation that, for different people, pornography performs different acts. For some, it is a form of entertainment, and for others, a form of subordination. In cases where there is a conflict between these perspectives, *who has the right to judge which act is being performed?* One could argue that women are in a better position to judge whether the act of subordination is performed since they are the ones who suffer the consequences. But such a line of argument would be inconclusive, as it pertains precisely to what is in dispute.

The third line of argument focuses on the fact that authority is one of the success conditions for verdictive acts. The crucial question then becomes: *do pornographic photographs possess the necessary authority to perform the act of subordination*? According to Langton, this question cannot be answered solely through philosophical analysis ("settled from the philosopher's armchair" (2009: 45)). What is necessary is to determine whether, for consumers of pornography, the discourse present in pornography indeed has the authority to perform the act of subordination. According to her, the answer is affirmative, as for many consumers, pornography instructs them on "which moves in the sexual game are legitimate" (2009: 45). For many men and young people, there is a monopoly of pornography as the sole source of information about the game of sex. There is abundant empirical evidence showing that frequent consumers of pornography are more likely to commit acts of violence against women because they feel legitimized to perform such acts by pornography.

Judith Butler (1997) criticizes MacKinnon's thesis of subordination as defended by Langton. According to Butler, MacKinnon and Langton's central mistake is attributing to pornography an almost *divine power*, one that makes women inferior simply by depicting them in a state of inferiority—similar to when God says, "Let there be light," and reality conforms to the command. The error, she argues, is "a kind of grammatical mistake in supposing that what pornography depicts women *as* can have implications for what a woman *is*" (Butler 1997: 68). However, Langton contends that the illocutionary act of ranking women as socially inferior makes them inferior in a *social sense*, but not directly in a *causal sense*.

To understand Langton's argument against Butler, it is important to note in what way verdictive acts are *self-fulfilling* in an illocutionary sense, as opposed to self-fulfilling in a perlocutionary or causal sense. As an example of a verdictive act, when a court declares a defendant guilty, they are considered guilty for the purposes of the court, even if they are causally innocent. Thus, the declaration of guilt is self-fulfilling in the sense that the defendant is *socially deemed guilty*, even if they did not actually commit the crime. It is at this point that Butler's analogy with divine power breaks down, as divine power would be self-fulfilling in a causal sense. In relation to pornography, the depiction of women as subordinate to men has the power to determine them as *socially subordinate* to men. This does not imply that women are *actually* subordinate, as if reality automatically conformed to their depiction. However, by declaring women to be *socially* inferior, pornography can also "make women actually become inferior, as they adapt to their social role" (Langton 2009: 109). The divergence with Butler lies in the fact that this does not attribute to pornography a magical, almost divine power to make reality immediately correspond to what is depicted. Instead, by influencing social perceptions, it can alter reality, with this alteration being a perlocutionary effect of the illocutionary act.

The second thesis defended by Langton is that pornography also performs the illocutionary act of *silencing* by preventing women from performing certain actions through their words. To defend this thesis, the author refers to Austin and distinguishes three modes of silencing. The first mode of silencing is to prevent the locutionary act from occurring. This can be achieved, for example, through intimidation or convincing someone that performing the locutionary act would be futile. Silencing, in this case, is the non-utterance of words. The second mode of silencing is related to the frustration of the perlocutionary effects of the speech. Langton offers the following examples: "one argues, but no one is persuaded; one invites, but nobody attends the party; one votes, hoping to oust the government, but one is outnumbered" (2009: 48). In these cases, the locutionary acts are performed, but the effects do not follow. The third mode of silencing is illocutionary silencing. Words may be spoken, performing the locutionary act, but the intended action does not take place. This is the case described by Austin as the failure of the speech act due to *illocutionary deficiency* (Langton 2009: 48).[7]

[7] In the article "Four Forms of Silencing: A Study Based on Rae Langton's Critique of Pornography" (da Silva 2023), published in Portuguese, I explore Langton's ideas as a starting point for constructing a *philosophy of silencing*. Extending speech act theory beyond Austin's perspective allows for the identification of an additional form of silencing by distinguishing between the utterance act and the locutionary act. Using John Searle's taxonomy of speech acts (1979), we can closely observe how the silencing of illocutionary acts is connected to the conditions required for the performance of these acts. This enables us to distinguish within what Langton refers to as illocutionary silencing three forms of silencing: the silencing of preparatory conditions, the silencing of sincerity conditions, and the silencing of essential conditions. In the article, I develop methods to prevent and combat these specific forms of silencing.

Langton's thesis is that pornography makes certain speech acts "unspeakable" for women. There are cases where women can utter the words (performing the locutionary act), and these words are understood, but there is an illocutionary deficiency—the intended speech act does not materialize. An example of illocutionary deficiency related to pornography is the refusal in sexual contexts. Often, a woman says "no," refusing to engage in sexual activity, and although the locutionary act is understood by the interlocutor (that "no" is literally said), the utterance is not taken as a refusal. In these cases, the *uptake* of the illocutionary act of refusal does not occur, even though the locutionary act of saying "no" is performed. According to Langton, in these cases, the illocutionary act of refusal becomes "unspeakable" for women. She can say the word "no," but this locutionary act is not taken as a refusal of the sexual act.

Illocutionary deficiency is distinct from perlocutionary deficiency. In the case of perlocutionary deficiency, the locutionary and illocutionary acts occur, such that the interlocutor understands that the performed act is a refusal, but still forces the sexual act. In this context, perlocutionary deficiency should be described as the rape of a woman who performed the illocutionary act of refusal. This differs from the case described in the previous paragraph where the utterance of "no" is not even taken as a refusal—although the violence inflicted is the same. Those who learn the conventions governing sexual interactions through pornography, which sexualizes the non-acceptance of refusal or the use of violence against women in response to refusal, are more likely to commit these two forms of violence (according to the empirical data).

One might say that, concerning the feminist *motto* "no means no," the two utterances of the word "no" serve distinct logical illocutionary functions. The motto aims to underscore that the locutionary act "no" should be recognized as the implementation of the illocutionary act of refusal—expressed in the second "no."

According to Austin's approach, refusal is an exercitive illocutionary act, which requires that the person performing the act has the necessary authority. If pornography denies women the power of refusal, it also strips them of authority, and in this case, authority over their own bodies.

Langton concludes that the speech act of pornography has the power to determine the success conditions of women's speech acts. Pornography determines the conditions under which women's words can or cannot successfully perform certain illocutionary acts and can or cannot produce certain perlocutionary effects. In the examples discussed, women are hindered from performing the speech act of refusal. Thus, Langton's analysis shows that the supposed right of men to produce and consume pornography also conflicts with women's freedom of expression.

3.3 John Searle's Taxonomy of Speech Acts

To deepen the understanding present in the use of John Austin's theory to pictures by David Novitz (1977) and Rae Langton (2009), it will be important to move beyond the taxonomy of speech acts proposed by Austin. John Austin (1962) was aware of the limitations of the taxonomy of speech acts he proposed. John Searle, in a chapter published in 1979, highlights some limitations of Austin's approach and offers a new way to classify illocutionary acts. In this chapter, I will focus on the criteria he proposed. The importance of this study lies in the fact that, in later chapters, I will use some of these criteria for a general classification of speech acts performed through the use of photographs and also for the construction of an ontology of photographic social practices.

Searle's starting point (1979: 2) is the critique of the metaphorical expression "illocutionary *force*," used by Austin as a way to characterize the speech act. On one hand, the idea of an *illocutionary force* seems to be a plausible way to approach illocutionary acts, since different acts can share the same propositional content as their basis and vary in terms of the strength of the speech act. The order to "close the door," the request to "close the door," the suggestion to "close the door" constitute different acts with different illocutionary forces that have in common the same propositional content. For this reason, Searle uses the formalization F(p) to characterize Austin's position, where "F" symbolizes the illocutionary force and "p" the proposition.

However, adopting F(p) for the analysis of speech acts can lead us to the misconception that "different illocutionary forces occupy different positions on a single continuum of force" (Searle 1979: 2). Although the difference between a *request* and an *order* can be thought of as a difference in *force*, the difference between an *order* and a *verdict* cannot be treated this way, as there are other differences to consider. They are distinct types of speech acts and not just different gradations on a continuum. Thus, understanding speech acts requires an understanding of the different dimensions of variation in speech acts, beyond their force.

Searle will classify the different dimensions of variation, in relation to which acts can differ (with force being only one of these dimensions). He addresses 12 dimensions of variation (1979: 2–8). Since some of them are not directly relevant to the studies conducted here, I will not explore them—our analysis will be restricted to seven categories. The first three are the most significant ones.

1. *Differences in the point (or purpose) of the (type of) act.* The purpose of the speech act is what is aimed at through the performance of the act. The purpose of an assertion is *to represent* how things are (whether true or false, accurate or inaccurate). The purpose of an order is for someone *to do* something. The purpose of a promise is *to commit* oneself to the interlocutor to perform an action. Different acts can share the same purpose, although they may differ from each other. An order and a request have the same illocutionary purpose (*for someone to do something*), but they are distinct speech acts. The purpose should not be confused with the perlocutionary effect of the act. The purpose is a characterization of the general type, while the perlocutionary effect is a non-necessary consequence that follows from its performance. More precisely, the purpose determines to which class of acts the illocutionary act belongs, and the perlocutionary act is a contingent consequence of the act.

2. *Differences in the direction of fit between words and the world.* Although speech acts are typically performed through a locutionary act, the relationship between language and the world within the illocutionary act can be conceived in different ways. The direction of fit aims to determine whether, in the act under analysis, it is the language that should fit the world or the world that should fit the language. When we assert that something is a fact, it is language that must fit the world—this fit being the way to determine the truth or falsity of the locutionary content. On the other hand, when an order is issued, it is the world that should fit the language—the fit of the world to the propositional content of the order being the measure of the fulfillment of the order. Thus, there are at least two directions of fit: word-to-world (as in the case of assertion)—which the author formalizes with the downward arrow symbol ↓—and *world-to-word* (as in the case of an order)—which the author formalizes with the upward arrow symbol ↑. However, not every speech act has a direction of fit. In the act of congratulating someone, for example, the speaker does not aim for their words to fit the world, nor for the world to fit their words. Thus, there is a third option of direction of fit, the null direction, which he formalizes with the symbol ∅.[8] As we will see next, there is also a fourth possibility in the case of declaratory acts, in which the utterance has the power to alter the world. A successful appointment to a position, for instance, has the power to give someone a new social or institutional status. For these cases, according to Searle, there is a double direction of fit—where, simultaneously, the language fits the world and the world fits the language. The author formalizes this using the symbol ↕.
3. *Differences in expressed psychological states.* Different speech acts presuppose different psychological states. For example, someone who asserts, describes, or reports expresses a *belief* about a certain propositional content; someone who orders, commands, or pleads expresses a *desire* for someone to do something; someone who promises or takes an oath expresses an *intention* to do something. Believing, desiring, and intending are different psychological states that are presupposed in these acts. In many cases, the psychological state constitutes one of the *felicity conditions* of the act. In Austin's terminology, the mismatch between the psychological states of the person performing the act and the states presupposed in the act would render the act *empty*. A promise made without the intention of fulfilling it, although it may be accepted by the interlocutor (resulting in the implementation of the act), would be an empty promise.
4. *Differences in the force or strength with which the illocutionary point is presented.* A request and an order, although they may have the same *purpose* (that someone does something), the same *direction of fit* (world-to-word), and may express the *same psychological state* (the *desire* that someone does some-

[8] In his 2010 book, Searle revises his terminology, conceiving that cases of null direction of fit can be understood as cases where the fit is already presupposed. If I congratulate someone for winning the race, winning the race is something presupposed in my speech act. For this reason, the author suggests replacing the symbol ∅ with the term "Presup" (2010: 29). However, the idea of a null direction of fit will be important for our understanding of the fictional uses of photographic images.

thing), differ in the force with which the act is performed—an order being a stronger act than a request. Thus, force can be conceived as a dimension of variation among some acts with the same purpose.

5. *Differences in the status or position of the speaker and hearer as these bear on the illocutionary force of the utterance.* As seen in our previous analysis of Austin (Sect. 3.1) and Langton (Sect. 3.2.2), certain speech acts have as a felicity condition that the speaker possesses the necessary authority to perform the act. Authority is generally related to the role played by the speaker in the society or institution where the act is performed (cases in which we could speak of a social or institutional hierarchy). Thus, the status of the speaker determines, in part, which acts are accessible to them and which acts are prohibited. For example, a soldier does not have the necessary authority to issue an order to an admiral. The locutionary act performed by a speaker who lacks the required status for the act is incapable of bringing about the intended act. Thus, in the case of an order, it is necessary that both the speaker and the hearer have the correct institutional positions for the implementation of the speech act.
6. *Differences between those acts that require extra-linguistic institutions for their performance and those that do not.* Some speech acts can only be performed when associated with an institution that is a prerequisite for the act. For instance, judgments, baptisms, and appointments require the existence of certain extralinguistic institutions for the utterance to result in the act's uptake. A judgment must take place in a court of law to yield a legally valid verdict. A baptism must be conducted within the context of a church. An appointment must be linked to an institution where the assigned position exists. These extralinguistic institutions generally have a hierarchy of roles, which determines the speaker's status or position (as described in condition 5). However, some authoritative speech acts do not rely on the existence of extralinguistic institutions and can, in principle, be performed by anyone in the appropriate circumstances. In these cases, the speaker's status required for the act does not depend on their position within a specific institution but on non-institutional contextual factors (as in the case of someone with a gun pointing at you who performs the authoritative act of issuing the order, "Give me your wallet").
7. *Differences in the style of performance of the illocutionary act.* Many speech acts share the same illocutionary purpose, direction of fit, force, institutional requirements, and speaker status, yet they differ in the style of the act's execution. Searle offers the following examples. According to him, "announce," "hint," and "confide" differ only in the manner in which the act is performed. In this sense, a difference in style constitutes a dimension of variation for acts that fulfill the same requirements in relation to the other dimensions of variation.

The complexity of how these different dimensions of variation in speech acts can be articulated is simplified by Searle by retracing these dimensions to four categories of conditions that may be required for the act to be implemented (1969: 64–71):

- *Essential conditions*: these require that the illocutionary purpose of the act be intended by the speaker.

- *Preparatory conditions*: these must be met for the act to be performed, such as the speaker's status and the existence of the extralinguistic institution.
- *Sincerity conditions*: these pertain to the possession of the necessary psychological states for the performance of the act.
- *Content conditions*: these ensure that the locutionary act performed is appropriate so that, in that context, with the fulfillment of the other requirements, the illocutionary act is implemented.

Based on these dimensions of variation and conditions, Searle's proposal is as follows:

> What I propose to do is take illocutionary point, and its corollaries, direction of fit and expressed sincerity conditions, as the basis for constructing a classification. In such a classification, other features (...) will fall into their appropriate places. (1979: 12)

The fundamental idea is that different illocutionary acts have different purposes, which will have different conditions of satisfaction. A belief will have *truth conditions*, a desire will have *fulfillment conditions*, and an intention will have *carrying out conditions* (Searle 2010: 29). This determines a direction of fit and may require additional conditions related to other dimensions of variation.

The proposal of a taxonomy based on the dimensions of variation allows Searle to reduce the complexity of illocutionary act types to five categories:

> We tell people how things are (Assertives), we try to get them to do things (Directives), we commit ourselves to doing things (Commissives), we express our feelings and attitudes (Expressives), and we bring about changes in the world through our utterances (Declarations). (1979: viii)

Let us examine each of these categories of speech acts (this taxonomy is found in Searle (1979: 12–20).[9]

I. *Assertives*. In assertive speech acts, the illocutionary purpose is for the speaker to commit (to a greater or lesser degree) to the truth of the asserted propositional content. In these assertive acts, the speaker expresses the psychological state of *belief* in the truth of the proposition. The act has a *word-to-world* direction of fit since the truth or falsity of the asserted propositional content is determined by its correspondence with the facts. The act can occur with varying levels of strength. Searle formalizes assertive acts using Frege's judgment stroke "⊦," the letter "B" as the symbol for belief, the downward direction of fit arrow "↓", and "p" as the symbol for the asserted proposition. Assertive acts have the following form: ⊦ ↓ B (p).

II. *Directives*. In directive acts, the purpose is to get the hearer to perform an action intended by the speaker. Examples of directive acts include ordering, commanding, requesting, pleading, and so on. In these acts, the speaker expresses

[9] The importance of this taxonomy lies in the thesis I will later present: we perform these five categories of speech acts through the use of photographs, making these categories also the basis for the taxonomy of different photographic social practices, as they can be distinguished by their illocutionary purposes.

the psychological state of desire "D" for a certain action "A" to be performed by the hearer "H". In Searle's symbolism, the purpose of directive acts is indicated by the use of an exclamation mark "!" and the *world-to-word* direction of fit by an upward arrow "↑"—since it is intended that the world should come to be as stated. Thus, directive acts have the following form: ! ↑ D (H does A). There is the desire "D" for the hearer "H" to perform the action "A," making the world fit what is desired by the speaker.

III. *Commissives.* The purpose of commissive acts is for the speaker to commit to performing a certain future action or actions. Thus, the speaker expresses the psychological state of *intention* to perform a future action. The direction of fit is *world-to-word,* as the world must align with the content intended by the speaker "S" for the intention of the commissive act "C" to be fulfilled. Thus, commissive acts have the following form: C ↑ I (S does A)—where "I" denotes the psychological state of intention.

IV. *Expressives.* The illocutionary purpose of expressive acts is to express a certain psychological state, which is a felicity condition of the act, regarding a certain state of affairs. Examples of expressive acts include congratulating, offering condolences, thanking, welcoming, and so on. Expressive acts have no direction of fit, as they do not aim to make the world fit language or make language fit the world. In an expressive act, the speaker communicates to the hearer their possession of a certain psychological state concerning something that may have happened to the hearer or to the speaker themselves. In an apology, the speaker expresses their regret for having done something. In a congratulation, the speaker expresses their joy for an achievement of the hearer. Expressive acts are symbolized by the letter "E," "P" (uppercase) symbolizes the psychological state, and "property" is what is attributed to the speaker or the hearer. For example, it is possible to congratulate the hearer for winning the tournament—the tournament victory being what Searle characterizes as a "property" of the hearer. Thus, expressive acts have the following form: E ∅ (P) (S/H + property), in which the speaker expresses a psychological state "P" in relation to something that has happened (the "property"), which may pertain to either the speaker or the hearer.

V. *Declarations.* Declaratory acts are those in which the successful performance of the act brings the world into conformity with the content uttered. Examples of declaratory acts include appointments, baptisms, marriages, dismissals, and so on. Declaratory acts are typically associated with institutions as a felicity condition of the act and require certain authority. If the speaker fulfills the requirements that are the felicity conditions of the act of christening, by saying "I christen this ship 'Antônio Carlos Jobim'," the ship is thus named "Antônio Carlos Jobim." Similarly, in appointing someone to a position, the successful act elevates the person to the assigned role, altering their status. The performance of the act thus has the power to alter the social or institutional status of the objects or people involved. Consequently, declaratory acts have a direction of fit that is not simply *word-to-world* or *world-to-word.* The declaratory act is an utterance that has the power to change the world. Searle represents this *sui generis* condition of declaratory acts with the symbol "↕". One notable aspect is that declaratory acts, being generally conventional procedures related to institutions, do not require psychological states as a felicity condition. For the act to be performed, it suffices that the correct procedure is followed, in the appropriate institution, by the appropriate people. Thus, declaratory acts ("D") have the following form: D ↕ ∅ (p). The symbol "∅" indicates the absence of a psychological state as a felicity condition for the performance of the act.

Another important contribution of Searle (1979: 30–57) is his distinction between *direct* and *indirect* speech acts. In direct speech acts, the speaker intends to perform a specific speech act, and the intended act is realized through the hearer's recognition of the speaker's manifest intention to produce such an effect. There is a dual intention (a Gricean *reflexive intention*): to perform the speech act and to have the act implemented by the hearer's recognition of the manifest intention. The hearer's recognition of the speaker's intention occurs due to the hearer's knowledge of the rules governing the utterance of the locutionary act. For example, when uttering the words "I apologize for my tardiness," the hearer's knowledge of the rules of the English language allows them to recognize that the speaker intends to make an apology. However, there are cases where the meaning of the uttered sentence and the intended speech act can diverge. *Indirect speech* acts are special cases of this divergence, in which the speech act performed is not grammatically related directly to the locutionary act. For instance, when asking the hearer at a dinner, "Can you reach the salt?", the speaker is not merely asking the hearer to assess their distance to the salt and respond if they can reach it, but also requesting that, if they can, they should pass the salt. Thus, in addition to the initial question, which requires an assessment and response based on their knowledge of the rules of the English language, the speaker is also making a request (not explicitly verbalized) for the hearer to pass the salt. This request constitutes an *indirect speech act.*

The difficulty posed by indirect speech acts is that the speaker's intention is not directly expressed in the content of the locutionary act (the words used). Thus, the literal meaning of the uttered sentence (the locutionary act) differs from the speaker's *intended meaning.* In the previous example, the literal meaning of the sentence is a question, though the speaker intends for the hearer to perform a specific action. According to Searle:

> In indirect speech acts the speaker communicates to the hearer more than he actually says by way of relying on their mutually shared background information, both linguistic and nonlinguistic, together with the general powers of rationality and inference on the part of the hearer. (1979: 31–32)

Thus, the performance of indirect acts presupposes, in addition to the theory of speech acts, an understanding of the background knowledge necessary for inferring the indirect act, which includes general principles of cooperative conversation (as outlined by Paul Grice (1957), which we will address in Sect. 3.6).

3.4 Different Ways of Connecting Photographs to Speech Acts

To thoroughly explore the pictorial and photographic applications of John Searle's contributions to speech act theory, we must first examine the various ways in which a photograph can be connected to speech acts. This analysis will enable us to classify five distinct ways of connecting photographs to speech acts, with the interpretive benefits being the distinction between *direct* and *indirect* pictorial photographic illocutionary acts and the possibility of distinguishing between cases where the role of the photograph is merely that of determining propositional content and cases where we could attribute to the picture, in a contextualized use, the determination of the illocutionary force.[10] Some of these connections will be relevant and will be the focus of our analyses in the following chapters, while others should be noted to be set aside.

One way a photograph can be connected to a speech act is through the idea that the photograph can *depict* someone performing a speech act. Consider, for example, a photograph depicting a judge in the act of striking a wooden block with a gavel, mouth open as if speaking. This picture can be described as a photograph of the judge performing the speech act of convicting the defendant. In this case, the photographic image portrays a person performing an illocutionary act. This instance could be characterized as a photograph *of* a speech act.

Another way a photograph can be connected to a speech act is through its connection to spoken or written language. We can point to a photograph and verbally apply different illocutionary forces to it. We can point to a photograph and say:

- "Do this."
- "I promise to do this."
- "This is what happened."
- "Bring this here."
- "This has excellent quality."

In this second scenario, the photograph primarily functions in a *propositional capacity*, determining descriptive content which, in contextualized use, contributes to specifying a proposition to which an illocutionary force is verbally applied. We can say that, in this case, the photograph is *part of* a speech act ("embedded in a larger communicative act" (Kulvicki 2020: 48)), with the communicative act being partly verbal. In this case, the illocutionary force may be explicitly applied (verbally) or presupposed by the contextualized use (with this idea being a significant element of the ontology of photographic social practices, which I will propose in Chap. 4).

[10] This analysis does not aim to be exhaustive but to establish a starting point for future studies. If other ways of connecting photography to speech acts are conceived in the future, they may be added to those examined here.

A third way a photograph can be connected to a speech act is by serving as a conventional element within speech acts. Consider a scenario where two individuals agree on a specific protocol: the placement of a photographic picture of a dog on a window signals the second person to close a door. In this case, the photograph is part of the command to close the door. Thus, in this third example, the photograph could also be described as *part of* a speech act.

However, a significant distinction exists between the second and third cases, even though in both scenarios the photograph is *part of* a speech act. In the second case, the photograph's role in the speech act is not arbitrary. When pointing to a photograph while saying "do this," the image's pictorial content and its denotative relationship to the world play a critical role in the performed act. Conversely, in the third case, any photograph could be transformed into a conventional sign, used as part of the command to close the door. This delineation allows us to distinguish between two ways in which photographs can be *part of* speech acts: as an integral component of the act (as in the second case) and as a sign conventionally related to the act (as in the third case).

The case in which a photograph serves as an essential component of a speech act can be classified as a *pictorial representation*, in the sense attributed by David Novitz (as seen in Sect. 3.2.1). According to the author, when pictures are used as pictorial representations, "pictures are used to convey information which pertains to, or is about, whatever is pictured" (Novitz 1977: 68). This idea can be expanded to include cases where the photograph is an essential component of a speech act in the following terms: we can add an illocutionary force to the photograph verbally, with the proposition to which the force is applied being "information which pertains to, or is about, whatever is pictured."

The fourth case is the most relevant to our purposes and occurs when the way something is portrayed in a photograph and how the photograph is used determine the illocutionary force of a speech act. For example, drawing from Rae Langton's analysis (as seen in Sect. 3.2.2), pornographic photographs are argued to possess the power to *rank* women as socially inferior. These photographs not only depict women in a subordinate position (as a locutionary act) or perpetuate acts of subordination (as a perlocutionary effect) but are, according to Langton, themselves illocutionary acts of subordination. Her key and groundbreaking assumption is that a photograph does more than determine pictorial/propositional content (depicting a certain scene) or produce some perlocutionary effects; it can also establish the illocutionary force of a speech act, depending on how it is used. In such cases, we can derivatively attribute to the photograph the realization of the speech act.

There are many instances that demonstrate the illocutionary potency of photography, some of which I will explore in the upcoming chapters. For example, the way food and objects are portrayed in advertising photographs can perform the directive speech act of prompting the spectator to purchase the product. We can criticize or praise someone depending on how they are portrayed in a photograph. We could classify this fourth way of connecting photography to speech acts as instances where the photograph itself performs the speech act. I will refer to this as the case of photography *as a speech act*.

There is still a fifth way in which photographs can be connected to a speech act. Often, the communicative intention is not directly tied to the pictorial content or expressive properties of the image, but rather to acts indirectly performed by its use or production. Consider the following example. On social media, it is commonly accepted that photographs form part of first-person narratives, and posting a photograph is often assumed to be temporally close to its production. We also assume that a photograph produced using traditional methods permits the inference that the photographer was physically present at the depicted scene when the image was taken. Therefore, when someone posts a photograph of a sunny beach on social media, even if they are not part of the pictorial content, they can be indirectly asserting their presence in that location, at that time, and enjoying the scene. These instances can be categorized as *indirect photographic speech acts*, based on inferences drawn from the assumption that the photograph was produced in a certain way and used with a certain purpose (in accordance with the specific social practice). We could also approach this idea of indirect pictorial speech acts using Grice's (1957) notion of *implicature*, developing the concept of a photographic implicature (as we will see in Sect. 3.6).

We can thus conclude that there exist at least five distinct ways in which photography can be connected to speech acts.

(i) The photograph *of* a speech act (depicting someone performing an illocutionary act).
(ii) Photography as an *integral part* of a speech act.
(iii) Photography as a *conventional element* part of a speech act.
(iv) Photography *as a speech act.*
(v) Photography as an *indirect* speech act.

These five modes are not mutually exclusive and can coexist and occur simultaneously. For instance, a photograph depicting a judge convicting a defendant (the photograph *of* a speech act) may portray the judge as aggressive and contentious, thereby casting doubt on their impartiality (photography *as a speech act*). Additionally, this photograph could form an integral component of a verbal critique directed at the judge by the user of the image (the photograph as an *integral part* of a speech act). Similar overlaps can occur with the other modes. The complexity inherent in communication through photography partly stems from this potential for simultaneous performance of different modes of speech acts. Moreover, a single use of photography may perform multiple speech acts of the same mode, as in the example of pornography which, according to Langton (2009), can subordinate and silence women.

While all five ways in which photography relates to speech acts are possible, our analysis will primarily focus on photography *as a speech act* and as an *integral part* of a speech act. We will also pay some attention to the case of *indirect speech* acts. The use of photography as a conventional element within a speech act does not necessitate further analysis beyond what is already established in speech act theories. Therefore, we will not dedicate ourselves to this case. The use of photography to portray people performing speech acts can be elucidated through elements present in picture theories—rendering detailed analysis also unnecessary.

To further examine our priority cases, we can employ the formalization proposed by John Searle of John Austin's speech act theory (Searle 1979). Different speech acts can be performed based on the same proposition. For instance, one might command "close the door," ask to "close the door," or promise to "close the door." The proposition "close the door" is the *base proposition*, and the speech acts of commanding, asking, suggesting, promising are the *illocutionary forces*. All these acts share the same proposition "p" and vary in illocutionary forces—symbolized by Searle as "F"—forming the scheme "F(p)."

When a photograph is an *integral part of* a speech act, its logical role is related to the propositional content. In such instances, a speech act (which can be verbal) is performed using the photographic image to determine the content or part of the propositional content of the act. For example, a photographic picture of a damaged vehicle could be used to assert that an accident occurred, to request help, to blackmail someone, or to claim compensation, among other possibilities. In this context, the photograph relates to the propositional content "p" in the speech act of the form "F(p)."

Regarding indirect speech acts, the logical function of the photograph may also be linked to the proposition that forms the basis of the act. However, unlike the aforementioned scenario, it uses an inferred proposition related to the picture as the basis for the act. For instance, posting a recent photograph of a landscape on social media may implicitly assert, "I am now at this location," even if I am not depicted. The inferred proposition, based on assumptions about the production of the image, becomes part of the base proposition "p" in the scheme "F(p)" (applying an assertive illocutionary force to the propositional content).[11]

But, in some cases, the illocutionary force does not need to be verbally applied to the image. For example, when a photograph is used as part of a restaurant menu, there is an implicit promise (which we will later describe as a commissive act) that if ordering that dish the restaurant will produce a meal similar to what is shown in the photograph. Here, the illocutionary force of the promise is indirectly linked to the image. It is part of the contextual rules of that social practice that there is a commitment for the establishment to provide a product similar to the one depicted. This commissive act is so central and important to this social practice of using photographs that breaking the promise can even lead to legal action and compensation (if the dish produced by the restaurant does not adequately match the photograph). The point to note is that the illocutionary force in this case does not need to be verbally applied but is presupposed within the context of use.

The philosophically more intriguing cases are those in which the photographic picture is not only related to the propositional content that forms the basis of the

[11] In this case, we could use Robert Hopkins' (2015) idea that what is communicated depends on the *means of production*. It is our belief that photographers are present at the moment the photographs are taken (along with other implicatures) which allows us to interpret the use of the photographic image as indirectly asserting, "I am now at this location."

speech act but also to the *illocutionary force* due to the way something is portrayed and how it is used. In this case, the role of the photograph is not only connected to the proposition "p" (determining the referent and its visual properties) but to the illocutionary force "F," composing the speech act "F(p)"—in which the photograph could play this dual role.

In some of these modes, we will be able to attribute the performance of the speech act to the photograph itself. However, since pictures are not individuals that perform speech acts, attributing the act to the picture itself would be derivative in nature, depending on someone's use of the picture. Thus, we have two possible theoretical paths: attributing the performance of the act to the user or producer of the picture, or, derivatively, to the picture itself. I do not believe this difference should be settled *a priori*, as if we had to choose which of these approaches was the correct one. The position I advocate is that in some uses, it is appropriate to attribute the act to the user or producer of the picture, while in other cases, the attribution should be made to the picture itself (particularly in cases where the illocutionary force is determined as part of the photographic social practice).

Faced with those possibilities, the central question becomes: how can photographic images perform speech acts, not only determining the base propositional content but also the illocutionary force of the act? Answering this question and related issues will be a central part of the remaining chapters of this book.

3.5 A Taxonomy of Photographic Illocutionary Acts

We have seen in the works of David Novitz (1977) and Rae Langton (2009) that photographs can be used to perform assertive, verdictive, and exercitive illocutionary acts. What other speech acts, then, can photographs be used to perform—especially in cases where we might ascribe the act to the photograph itself, albeit in a derivative sense? This question leads us directly to the problem of classifying speech acts.

The classification of speech acts is a complex issue. Austin (1962) proposed that speech acts are innumerable, a notion echoed by Wittgenstein (1953), whose ideas suggested the existence of countless language games. As seen in Sect. 3.3, John Searle offers an elegant solution to this problem by retracing all speech acts to five distinct general types. According to Searle:

> We tell people how things are (*Assertives*), we try to get them to do things (*Directives*), we commit ourselves to doing things (*Commissives*), we express our feelings and attitudes (*Expressives*), and we bring about changes in the world through our utterances (*Declarations*). (1979: viii)

To develop a foundational taxonomy of photographic speech acts, which could serve as a starting point for future refinements, it is essential to evaluate whether photographs possess the capacity to perform these five general types of speech acts. I believe the answer to this matter is affirmative and that all five types of speech acts can be performed using photographs. We may consider these speech acts as the five general types of *photographic illocutionary acts*. It is also important to consider whether, in the execution of these five types of speech acts, the logical role played by photographs is limited to determining propositional content or if it extends to determining the illocutionary force. With this, I hope to provide a starting point for future works to refine these categories, determine their subgroups, and identify more nuanced modalities (a task that will not be undertaken in this book).

For the development of this foundational taxonomy of photographic speech acts, I will base my analysis on the work done in Sect. 3.3, which examines John Searle's theory.

3.5.1 Assertive Photographic Acts

In assertive speech acts, the illocutionary purpose is to assert something about the world in a manner that commits the speaker to the truth of the propositional content determined through the utterance act. The psychological state expressed in assertive acts is one of *belief* (which may vary in degrees) concerning what is asserted. Additionally, assertive acts have a *word-to-world* direction of fit, since it is the correspondence between what is said and the world that determines the truth or falsity of what is asserted.

According to Novitz (1977), this is a speech act common to picture use. We frequently employ pictures to refer to objects and attribute properties to them, committing ourselves to the truth of the propositional content expressed through the pictorial content. Aligned with Novitz's thesis, which applies to pictures in general, I believe it would be uncontroversial to state that we often *use* photographic images with a commitment to the truthfulness of the descriptive/pictorial content, thereby engaging in assertive speech acts. The fact that we label photographs as true or false attests to their assertive use. This would imply a *picture-to-world* direction of fit. In the basic use of photographic images for assertive acts, the pictorial content of the image establishes a propositional content, or part of a propositional content, and in its contextualized use, there is a commitment regarding the truthfulness of this content. In assertive photographic speech acts, whether the photograph corresponds to the world determines its truth or falsity.

Such assertive use of photographs is so deeply embedded in Orthodox theories that some authors assume every photographic image inherently performs such an act (as something that would follow from the way it is causally connected to the referent). For instance, Roland Barthes (1981) encapsulated the supposed essence of photography in the idea that every photograph asserts "*that-has-been*." The assertive nature of photography is tied to the emphasis on the causal mechanism that fixes the pictorial reference, thereby connecting the pictorial content with a *picture-to-world* direction of fit "↓" (as if the photograph were constantly pointing to its causal referent).[12] Thus, the assertive illocutionary force of photographs, in an Orthodox Barthesian perspective, would be inherent to the photograph itself. Consequently, it could be argued that in assertive speech acts, photography not only determines the proposition that serves as the foundation of the act (the asserted proposition), but, due to its causal genesis (the means of production), it also plays a role in determining the illocutionary force of the act.

However, as analyzed in Sects. 1.5 and 1.6, the Orthodox view does not encompass the essence of all photographic practices. When considering the speech act and the direction of fit, the fundamental point is not the metaphysical issue of whether photographs are inherently capable of truth or falsity or causally connected to the denoted object. Instead, the focus is on *how we use* photographs in communicative contexts, whether we assume the picture should conform to the world or the world should conform to the picture, including possibilities of null or dual direction fits (as I shall argue later). The speech act and its corresponding direction of fit pertain to the communicational use of the picture and the speech act performed through its contextual use. Thus, regardless of photographs having a causal genesis that links them to their referents, not every communicative use of photographs is assertive.

If we acknowledge the possibility of fictional uses of photography, as supported by Atencia-Linares (2012), it becomes evident that certain uses do not aim to depict the world or establish models for it. Such fictional uses can be understood as having a null direction of fit "∅". Additionally, as we will see shortly, in directive and commissive acts, the picture is used with a *world-to-picture* direction of fit, and declaratory uses would have a double direction of fit.[13]

Photographs are employed for assertive speech acts in situations like (but not limited to):

- Description: A photograph can be used to describe a scene, object, or person.
- Documentation: A photograph can serve as documentary proof of a phenomenon or event.
- Testimony: A photograph can provide testimony for a specific claim.
- Evidence: A photograph can serve as evidence in legal proceedings.

These uses constitute assertive acts when the user believes, or is presumed in the context of use to believe, that the propositional content determined by the

[12] From Barthes' Orthodox perspective, we could say that photography would be the language for which Cratylus waited for more than 2000 years, in his attempt to point to the present that constantly vanishes in the flow of time. Photography has the power to point to that which, in the river of time, remains unchanging, constantly receding into the past. In the article "What does a presentist see when she looks at photographs of dead relatives?" (da Silva 2018), I explore how the Orthodox approach of photography presupposes a form of metaphysical realism of the past, which could not be sustained by a presentist.

[13] I will argue later that many speech acts can be performed simultaneously through the use of the same photograph. Thus, there would be no logical impediment to the same photograph being used as part of a directive act while also serving assertive, commissive, expressive, or declaratory purposes, each with its own distinct direction of fit relative to the specific act.

descriptive/pictorial content conveys information about the world that can be evaluated in terms of truth or falsity.

As explored in previous chapters, the propositional content determined by a picture is contingent on contextual elements shared between the user and the viewer. An image can give rise to multiple propositions, and the determination of which entity is being referenced, as well as the properties attributed in the picture, hinges on the background information and conceptual framework shared by the user and viewer. This contextual determination of the propositional content is critical in assertive speech acts, as it is this pictorial propositional content that the user, or derivatively the picture, asserts about the world. This intersection highlights a point of contact between *near-side* and *far-side* pictorial pragmatics (a distinction introduced at the beginning of Chap. 2). It is the contextualized use that establishes the context-sensitive factors determining which proposition is asserted through the intentionally assertive use.

In assertive speech acts, if a photograph does not correspond to the world, it is considered false; conversely, if it does correspond, it is deemed true. I label those scenarios as *pictorial truth* or *pictorial falsity*. In those cases, the truthfulness or falsehood is ascribed to the image's pictorial content (constituting the *locutionary act*).

The determination of truth or falsity will hinge upon which type of visual information present in the photograph holds significance for the social practice in which the photograph is used, and to what extent the practice imposes accuracy criteria for establishing the correctness or incorrectness in depicting the world. For example, one context of use may allow chromatic deviation but not geometric distortions, while another context may allow geometric distortions but not chromatic deviations. Thus, a photograph with chromatic deviation but no geometric distortions would be considered true in the first context of use and false in the second. Contexts with high epistemic rigor (such as the use of photographs in science and as a tool for medical imaging diagnosis) generally have strict criteria for accuracy and veracity—along with verification methods to determine truth or falsity. A picture that, in a permissive context with few constitutive rules, may be considered true could be deemed false in a restrictive context with many constitutive rules. (Constitutive rules will be explored in greater detail later in Sect. 4.2).

An assertive speech act can also be performed indirectly using photographs. If in the contextualized use of the picture what is being asserted is "I am (or was) here at this location," even if I am not depicted (but inferred from the fact that I took the photograph at that location), this indirect assertion is subject to a judgment of truth or falsity.

A common philosophical misstep is the assumption that every assertive use of photographs intends to assert a proposition with a *singular content* fixed by the causal mechanism of pictorial reference. The error (already addressed in Sect. 2.3) lies in presuming that each assertive use of photographs seeks to assert a proposition about the causal referent (a singular content) and ascribe visual and/or expressive properties to it. While photographs are often used for such singular assertive acts, different assertive uses are possible. As seen in our analyses of reference theories

(Sects. 2.2 and 2.3), pictorial reference can be established descriptively, causally, iconographically, and intentionally. It is the contextualized use that determines which referential mechanism is linked to a specific assertive use—within the scope of *near-side pictorial pragmatics*. While the primary mechanism of photographic reference is often the causal connection, descriptive, iconographic, and intentional references can play important roles in the content asserted through the assertive speech act, allowing the use of photographic images to perform assertive acts with content that is not singular.

A common use of photographs in assertive contexts, where the photograph is employed to express a non-singular proposition, is through generalization (as previously analyzed in Sect. 2.3)—which, in Robert Hopkins' (2015) formulation, is the use of a photograph to exemplify a *type* (discussed in Sect. 1.2). For example, a photograph of an individual of a particular race, creed, gender, or nationality might be used to represent all (or some) members of that group, and the visual and expressive properties of the person in the photograph could be asserted as predicates for the entire class (or part of it). The crucial point is not to deny that this same photograph could be used with a singular content, but to recognize that the communicative intention may be to assert a general proposition about people of a particular race, creed, gender, or nationality through the picture's use. In this case, the determination of the truth or falsehood of the assertive illocutionary act should be made in relation to the general proposition asserted in the contextualized use, as this is what the user intends to assert.

This distinction allows for a bifurcation of truth in relation to the assessment of whether the pictorial content is true or false, and the truth or falsity of the assertive illocutionary act performed through the use of the photograph. Regarding assertive uses, when faced with a photograph, our initial tendency is to question whether the pictorial propositional content corresponds to the world. But while an image may accurately and genuinely depict a scene in a singular assertive use (attributing properties that are contextually accepted as accurately describing the causal referent), the communicative intention, if general, could perform an assertive illocutionary act with a different truth value. For instance, in linguistic terms, the singular proposition "João is lazy" might be true, whereas the general proposition "people of the class to which João belongs are lazy" could be false. If a photograph of João is used with the intention of making a general assertion about the class to which he belongs, even if the pictorial content is accurate (correctly representing João), the illocutionary act (the general assertion) could be false. Conversely, an image that inaccurately and falsely depicts João could be used to make a general assertive statement that is true.

Hence, although a photograph might be truthful from a pictorial standpoint (when evaluated in relation to the singular causal referent), the assertive speech act performed could be false, and vice versa. For example, a photograph of the Hitler Youth, showing them as strong and healthy, may be pictorially and causally accurate and genuine, respectively. However, if the contextualized use has the communicative intention of asserting that "the Aryan race is superior [as they are strong and healthy]," it would render the propositional content of what is depicted true (if assessed for its truthfulness), whereas the illocutionary act (the general proposition—what is asserted using the picture) would be false. *In other words, accurate and causally genuine photographs can be used to assert false propositions (and vice versa).* If the image is employed to convey the general assertive statement "the Aryan race is superior," the assertive speech act would be an instance of an *illocutionary falsehood.*

Those cases could also be analyzed as a dual assertion: one with singular content (related to the content fixed by the causal relation with the referent) and another with general content (stemming from the process of generalization). And generalization could also occur through the use of iconography, in which an element of the picture is converted into a symbol that denotes a class of objects. The lesson to be learned is that determining a photograph's truth or falsity involves distinguishing between locutionary and illocutionary acts and understanding the specific proposition being asserted, which could be of a general kind.

Furthermore, the bifurcation of locutionary and illocutionary truth allows for illocutionary truth to be independent of primary denotation. A fictional photographic picture, lacking primary denotation (in causal terms) and not asserting a *pictorial truth* about the world, can still be used for an assertive illocutionary act that is subject to truth or falsity. For instance, a photoshopped image with fictional content (recognized as a montage) can be employed to assert something that may be true or false. This scenario is typical in the case of *memes*, which are nowadays abundant on social media. Often, it is inappropriate to assess the pictorial content of these memes for truth or falsity, as they are commonly understood to be fictional montages. However, such fictional pictures can still be used to perform assertive speech acts, conveying illocutionary truths or falsities. A meme used to assert that "the Aryan race is superior" would be an illocutionary falsehood, even if its pictorial content is acknowledged as fictional and not subject to judgments of truth or falsity in photographic terms.

Another approach to distinguishing between two modes of determining truth in the use of pictures is through the use of metaphors, as proposed by Daisy Dixon (2018). In Dixon's theory of metaphor, the pictorial content of the image could determine a source domain (what the image portrays), and the contextualized *use* could have a target domain, in which the source domain is mapped. For example, the image of a lightbulb could signify technology, so a photograph depicting light as bringing clarity and warmth could be used to signify the enlightenment and security that technology affords us. Dixon's theory of metaphor could be interpreted in relation to the pictorial generalization or to iconographic reference, in which an element present in the pictorial content comes to be taken as a symbol denoting a certain object or objects by convention. In cases where the object of the source domain is conceptually related to the target domain (for example, belonging to a certain class of objects), the metaphor would be understood via generalization, and in arbitrary cases, via iconographic reference. The crucial point to note is that the use of metaphors also allows for the bifurcation of locutionary and illocutionary truth. Pictures can be pictorially true or false (if analyzed in relation to the correspondence of the pictorial content with the world) independently of the truth or falsity of the metaphorically asserted proposition—which itself could also be true or false. The determination of which proposition is asserted, whether singular or general, and what quantification is attributed to the general proposition (whether it pertains to some, one and only one, or all) will depend on the context of use and/or the communicative intention of the user.

3.5.2 Directive Photographic Acts

In directive speech acts, the illocutionary purpose is to prompt the listener to perform an action intended by the speaker. These acts are geared towards aligning the world with the speaker's words, thus having a *world-to-word* direction of fit, and often require a degree of authority for effective performance. Common forms of directive acts include orders, suggestions, advice, and requests.

It is reasonable to consider that a photographic picture performs a directive act when used with the intention of prompting the viewer to engage in a specific action.[14] There are numerous instances where the use of photographs can be associated with directive acts. Here are some examples (though this list is not exhaustive):

- Educational Use: A photograph in an educational poster can be used with the purpose of prompting the viewer to act in a certain way.
- Instruction: A photograph can function as an instruction to indicate how something should be used.
- Medical Use: A photograph can be used to encourage people to maintain healthy habits, such as regular exercise or a balanced diet.
- Blackmail: A photograph can be used to blackmail someone into performing a certain action.
- Advertising: A photograph in an advertisement can serve to prompt the viewer to purchase a particular product or service.

In most directive cases involving photographic images, the role of the picture extends beyond merely determining the propositional content of the speech act, to which a directive illocutionary force would be added (either verbally or presupposed within the context of use). The manner in which the picture depicts a subject can motivate individuals to undertake specific actions, such as behaving in a certain way, purchasing products, or using a particular piece of equipment. Consequently, we could ascribe to the picture itself, in a derivative sense, the directive speech act of prompting the viewer to engage in the desired action. If one prefers to assign the directive act to a specific agent, it could be attributed to either the photographer who created the image or the user who chooses to employ it.

Many directive acts rely on what Jerry Fodor refers to as "reflexive intention" (2012: 63). This concept, linked to Paul Grice's theory of communication (analyzed

[14] Dominic Lopes examines the concept of directive pictures in his 2004 paper.

in Sect. 1.3), posits that effective communication requires the speaker's communicative intention to be manifest and recognized by the interlocutor. For instance, for a verbal statement to successfully order an interlocutor, the latter must recognize the speaker's intent to issue an order.

Fodor (2012), in his analysis of Arthur Danto's (1981) theory of art, differentiates between uses of pictures that depend on reflexive intention and those that do not. This distinction helps to separate directive uses into what might be termed "Gricean cases" (reliant on reflexive intention) and "non-Gricean cases" (independent of reflexive intention). For instance, in the use of a photograph for blackmail, it is crucial for the viewer to recognize that they are being blackmailed through the use of the picture. This recognition of the user's intention is what constitutes the use of the image as an act of blackmail, qualifying it as a *Gricean case* where the photograph serves as a vehicle of communication under reflexive conditions (Fodor 2012: 62).

In contrast, there are non-Gricean directive uses of photographs where reflexive intention is not central or may even be intentionally obscured. A photograph can act as a guide, directing the viewer to perform certain actions without their recognition of the user's intent. Fodor notes an example of an advertisement that influences the viewer to buy a product in a "subliminal and associative" manner (2012: 63). In such instances, the photograph performs a directive speech act in a non-Gricean way, not requiring a reflexive intention. The picture harbors the directive intent to induce the viewer to purchase a product, achieving its aim through "hidden persuasion" (2012: 63). In other words, the user's intention is directive (aiming to lead the viewer to perform an action), although it does not constitute a case of genuine communication, as it is not based on the recognition of the communicative intention. In this case, we could claim, drawing on Grice's theory, that such use of photography would not constitute communication. However, we could understand non-Gricean directive uses of photographs as part of established social practices (such as advertisement photography) that constitute communicative contexts—where photographs are used by individuals with a certain intention related to an illocutionary act directed at viewers. And such uses could be understood and explained within the framework of a theory of pictorial communication. In other words, although these uses may not be genuine instances of communication, they could be explained within the scope of a theory of photographic communication.

Fodor emphasizes that distinguishing between Gricean and non-Gricean uses of pictures is pivotal in his effort to differentiate advertising from art, in the context of Arthur Danto's definition of art (1981). According to Fodor, the requirement for an object to be *interpretable* as a necessary condition for it to be classified as art inherently makes art Gricean for Danto. For something to be art, it must be intended and recognized as such. Conversely, advertising can achieve its directive objectives even if not explicitly recognized as advertisement, operating in a subliminal and associative manner as a form of hidden persuasion.[15]

In the terms of speech act theories of Austin (1962) and Searle (1969, 1979), we could interpret the distinction between Gricean and non-Gricean cases as involving a distinction between *illocutionary acts* and *perlocutionary acts*. When there is a communicative intention that must be made manifest by the user and recognized by the interlocutor, we properly have the performance of an *illocutionary act*. However, when this use does not rely on a reflexive intention, the intended effect would be a consequence of the picture's use, which could be understood as a *perlocutionary act* resulting from that use. Thus, Gricean cases would properly involve directive illocutionary acts, whereas in non-Gricean cases, the action prompted would be more accurately described as a perlocutionary effect.

A crucial aspect of directive acts is that they often necessitate a certain level of authority for correct performance, and this authoritative aspect may require ties to extralinguistic institutions. For example, in educational and medical contexts, when photographs are used for directive purposes, their authoritative aspect stems from the way the picture is linked to educational or medical institutions. Take, for example, the photographs on cigarette packs designed to deter smoking. In Brazil, these images are typically accompanied by information indicating that the warning comes from the Ministry of Health. This imbues the use of the image with a Gricean dimension, as the directive act is recognized as being intended by an authoritative institution.

Advertising photographs frequently capitalize on the element of authority as part of the pictorial content (and not merely as related to the authority of the user performing the act). Pictures featuring public figures endorsing a product tie the directive act to the authority represented by the depicted figure. The authority of the figure lends credibility and persuasive power to the speech act performed by the photograph. Companies are institutions, and it can be said that there exists a social hierarchy of corporate authority, with some companies recognized as having greater authority than others. The advertising photograph of a new smartphone from the most reputable brand on the market uses the institutional weight of the company to reinforce the directive act conveyed through the photograph, which can be interpreted as "buy this product."

However, as we have seen, there are instances where photographs can be used in directive contexts in a non-Gricean manner, where the need for authority and institutional connection may be less pronounced or even unnecessary. For example, even without explicit recognition that the Ministry of Health is the driving force behind the directive act to stop smoking, the photographs on cigarette packs could still achieve this directive purpose through hidden persuasion. This indicates that the same use can exhibit both Gricean and non-Gricean aspects. We could formulate

[15] My intention is not to suggest that art, as defined by Danto, is directive, but rather that directive acts may or may not require a reflexive intention.

this distinction as the performance of a *directive illocutionary act* that may also include *directive perlocutionary effects* (as hidden persuasion).

3.5.3 Commissive Photographic Acts

Commissive speech acts aim to express the speaker's commitment to a certain action. These acts manifest the speaker's psychological state of intention and are directed towards aligning the world with the uttered words, with a *world-to-word* direction of fit. When I say, "I promise we will go to Tuscany," I am expressing the intention to bring the world into alignment with the stated state of affairs. Promises, refusals, and oaths are examples of commissive acts.

It is possible to perform commissive acts through the use of photographic images. We could point to a photograph and perform the following acts:

- Promise: "I promise to do this."
- Refusal: "I will not do this."
- Oath: "I swear to do this."

In these cases, the photograph determines a propositional content to which the illocutionary force of the verbal commissive act is added. I posit, however, that photographs can also engage in commissive acts with the illocutionary force being related to the purpose of the photographic social practice.[16]

To grasp this concept, it is vital to differentiate commissive from assertive acts. In assertive acts, the user employs the picture with a psychological mode that might be construed as a *commitment* that the pictorial content fits the world (in basic cases of pictorial assertion). This could lead to the mistaken notion that there is a commissive aspect within assertive acts, as the user/producer is implicitly *committed* to the veracity of the visual information presented in the photograph (as if saying, "I promise that the picture is truthful"). However, this does not encapsulate what I aim to define as a photographic commissive act. The commissive element in photographic commissive illocutionary acts is not related to the correspondence between the pictorial content and what was depicted *at the time it was depicted* (in the *past*), but rather to a commitment directed towards the *future*, in which there is the presupposition that the world should correspond to the picture. In assertive uses, the direction of fit is *picture-to-world*, whereas in commissive uses, the direction of fit will be *world-to-picture*.

[16] I am grateful to Denis Perrin, James Openshaw and Kourken Michaelian (Université Grenoble Alpes, France) for highlighting the importance of commissive photographic speech acts. In my initial formulation of the theory, which I presented in a series of lectures at the Centre for Philosophy of Memory, Université Grenoble Alpes (France), in 2023, I disregarded commissive acts as being neither common nor central to photographic communication. Fortunately, the questions and critiques raised by Perrin, Openshaw, and Michaelian prompted me to rectify this oversight.

Take, for example, the use of a photograph on a restaurant menu. Beyond the directive act of prompting a customer to order the featured dish (whether as an illocutionary act or a perlocutionary effect), there is an implicit *promise* from the restaurant that the dish, if ordered, will resemble the picture. In this case, the photograph serves as a model that the real-world item is expected to match, having a *world-to-picture* direction of fit. If the restaurant provides something that does not resemble the photograph, we would say that the implicit promise has been broken. The restaurant's commitment to provide something similar to the photograph is so central to this social practice that a breach may even be subject to legal action and monetary compensation. In other words, there are contextualized uses of photographs as part of social practices in which the use performs a commissive act that introduces future obligations (which we will describe in Chap. 4 as *negative deontic powers*) related to the fit between the world and the picture.

There are also commissive uses of photographic images when they are employed with the commitment that a certain place or object possesses certain visual properties, not just in relation to the time the photograph was taken, but also towards the future. For example, when an establishment, like a hotel, uses a photograph in an advertisement, it is not merely performing an assertive act that the hotel appeared as such *at the moment* the picture was taken. There is an implicit commitment that if you visit that location in the near *future*, you will encounter something with those visual properties. If the location no longer resembles the photo (as after a renovation), the hotel would be legally required to change the photograph in its advertising—since it would not comply with advertising regulations (which we will later classify as part of *pictorial content conditions* for performing the illocutionary act (Sects. 4.2.4 and 4.3)).

We can thus summarize that photographs are employed in performing commissive acts whenever there is a *future-oriented* commitment related to the use of the picture, which is linked to its pictorial content or denotation, with a direction of fit from the *world-to-picture*.[17]

3.5.4 Expressive Photographic Acts

Expressive speech acts express a psychological state about a certain state of affairs. These acts do not have a direction of fit, since the expression of a mental state about a state of affairs presupposes the state of affairs (not requiring that the world fit the

[17] We can also locate the commissive dimension of the use of a photograph by following one of Rae Langton's lines of reasoning (2009). Although inference of the act from the effect may be fallible, we can trace the effects to the illocutionary act performed by conceiving the act as the cause of the effect. If the use of a picture can be described as misleading or disappointing with respect to the fit between the world and what the picture depicts (in a *future-oriented* sense), this may indicate that, to some extent, a commissive act was presupposed in the use of the picture. However, as Langton demonstrates (in the analysis we conducted in Sect. 3.2.2), this line of reasoning is fallible.

words or the words fit the world). According to Searle, examples of expressive performative acts include "I *apologize* for stepping on your toe," "I *congratulate* you on winning the race," "I *thank* you for giving me the money" (1979: 23—emphasis added). In Searle's examples, the verbs are associated with the expressed psychological states, and the subsequent content specifies the state of affairs to which the psychological state relates. These acts characteristically require the actual possession of the expressed psychological state as a *sincerity condition*. If the speaker does not possess this state, the act can still be performed but would be considered insincere or empty, such as an apology issued without genuine remorse.

Photographs, like pictures in general, possess significant expressive power. It is a widely accepted notion that photographs are capable not only of determining propositional contents but also of expressing mental states and emotions. Therefore, attributing the capacity to perform expressive speech acts via photographs may not be a controversial issue. However, it is important to note the extent to which using a photograph to perform an expressive illocutionary act differs from the mere expressiveness of the picture or its perlocutionary effects.

In Sect. 2.4, we explored Dominic Lopes' theory of pictorial expression (2005). We addressed how expressive content can be linked to *figure* expression (be they human, animal, or anthropomorphized objects), *scene* expression, and *design* expression—to which we also added the expressiveness of *content-neutral vehicle properties*. We also examined expression through natural, conventional, and *sui generis* mechanisms (for instance, pain shown on a human face is a natural expression, a thumbs-up gesture conventionally expresses contentment, and keying someone's car is a *sui generis* expression of anger). Photographs can use these various means to express mental states, feelings, and emotions.

The main point of Lopes' analysis was understanding how scene and design can be expressive even in the absence of figures. Of the interpretative perspectives he explored, three are especially pertinent to the idea of illocutionary expressive speech acts. In the perspective called *personalism*, the mental state expressed by the picture is attributed to the producer of the picture, reflecting the common belief that a picture expresses the artist's feelings. Alternatively, *arousalism* posits that the bearer of the mental state expressed by the picture is the spectator; thus, a picture could be said to express sadness if it induces sadness in those who view it. Lopes proposes the impersonal theory of pictorial expression, in which the picture can have an *expression-look* that conveys a certain mental state, without the need to attribute this to a subject (whether the producer or the viewer).

But there is an important difference between the expressiveness of the picture and the use of the picture for performing what I call an *expressive photographic illocutionary act*. The expressive act is achieved through *reflexive intention*. The use of the photograph's expressiveness must be intentionally manifested, and this intentionality must be recognized by the spectator. In other words, the picture is not just expressive; there is a manifest intention that the use of the picture is related to this expressiveness, and this must be recognized by the viewer. The indication that the use of expressive properties is intentional can also be part of the social practice, such as in certain artistic contexts where the function of the practice and the purpose of the photographs are related to their expressive properties. In these cases, the artistic value of the picture would be associated with the photographer's ability to perform expressive speech acts, and this is something recognized by users and spectators participating in the practice.

This allows us to draw an important distinction regarding David Novitz's theory (1977). As seen in the analysis of Novitz's theory in Sect. 3.2.1, according to him (in a previously cited passage): "expressive properties are ascribed to pictures independently of how they are used [...] *never* in virtue of their use in an illocutionary act" (1977: 150, emphasis added). The point of convergence between the position I adopt and Novitz's could be expressed by *partially agreeing* with the idea that expressive properties *can be* ascribed to pictures independently of how they are used. Photographs can be considered expressive even if they are not intentionally used with a communicative purpose. However, the point where we diverge is that, in the perspective I adopt, expressive properties *can be part of an expressive illocutionary act* when the picture is used with an overtly communicative intention that employs the expressiveness of the picture. The expressiveness of the picture can be recruited as part of a speech act in which the user expresses their *psychological state* about the depicted state of affairs. Novitz does not consider this possibility since his analysis is limited to assertive acts.

Let us consider an example. We might use a photograph in which the scene has a certain expressive property with the communicative intention of showing how we felt regarding what is depicted, even if we are not depicted in the scene. The expressiveness of the picture is part of the expressive act that conveys *my psychological state* about the depicted *state of affairs*. In this case, correctly understanding what the use of the picture aims to communicate requires the viewer to attribute a *personalist* approach to pictorial expression, where the focus of attribution of pictorial expression is the image producer or user (as discussed in Sect. 2.4). The recognition of the communicative intention by the viewer requires acknowledging the user's intention to use the picture's expressiveness to express their psychological state. In this case, the image is not merely expressive (in Novitz's sense) but is used to perform an expressive illocutionary act.

We can further explore the example above to show how, in such cases, the possession of the psychological state could also be taken as a *sincerity condition* for the act's felicity (in the technical sense of John Austin (1962) and John Searle (1969, 1979), which we analyzed in Sects. 3.1 and 3.3). If I use a photograph with the manifest communicative intention that the expressiveness of the picture conveys my psychological state about what is depicted, and if viewers later discover that this was not my psychological state, they may well feel that in some respect I deceived them. For example, if I use a photograph in which the scene expresses happiness as part of a first-person narrative, as an expression of my happiness, but I am actually not happy about that, my interlocutors will likely perceive such use as empty or deceitful.

Certainly, in many cases, determining whether someone possesses a certain psychological state to fulfill the sincerity condition of the act is not an easy task. The attribution of mental states frequently hinges on the person's behavior, which can indicate the presence of such mental states (such as the behaviors of pain, joy, or sadness). However, not all mental states manifest themselves in this way. In many cases, what is often taken as a condition of sincerity is a biography that aligns with the presence of such psychological states. The actions performed by someone and the choices they make throughout their life provide, with a certain degree of precision, indications of their mental states, as part of the reasons and causes for those actions and choices. Thus, fulfilling the sincerity condition often becomes a matter of determining whether the user's biography is consistent with having the required states.

Although the distinction between the expressiveness of the picture and the performance of expressive illocutionary acts is subtle and depends on various context-sensitive factors, it is important to note that, in pictorial expressive acts, the picture not only expresses the mental state but also can determine the state of affairs with which this mental state is concerned, encompassing both the proposition "p" and the illocutionary force "F" in the scheme "F(p)." The photograph assumes a dual role, relating to both the illocutionary force and the propositional content upon which this force operates, since the expressiveness of what is depicted is recruited as part of the illocutionary force. For example, a beach photographed in black and white, with shades of gray, could be used to express my melancholy about the depicted location. The same location captured in vibrant colors and bright light might be used to express joy and vivacity. The mental state I express is thus part of the pictorial content.

The importance of expressive speech acts through photographs is so significant that various aesthetic and artistic practices are associated with the photographer's ability to perform expressive acts. In these cases, the function of the social practice can be directly related to the production and use of photographs in which expressive acts are performed by photographers and artists. Within these practices, it is often assumed that the more expressively potent the picture, the greater its aesthetic or artistic value. And expressive practices can vary regarding different modes of expressiveness (figures, the scene, the design, and/or the content-neutral properties of expression), so that certain expressive practices may prioritize specific modes and require an understanding of this restriction for engagement, both from producers and from viewers.

But, once a certain type of expressiveness becomes the aesthetic standard of a social practice (functioning as a constitutive rule), the importance of having the mental state as a sincerity condition diminishes, and what takes centrality is whether the picture has the *expression-look*—a configuration that, in certain circumstances, serves to indicate the emotion (as analyzed in Sect. 2.4). The artistic practice is thus structured around the participants' effort to match the expressiveness that constitutes the aesthetic standard of the practice, independently of whether they possess the mental state expressed.

Nevertheless, the expressive use of photographs may not only relate to the artist's ability to perform expressive acts but also to the picture's capacity to evoke certain mental states in the viewer. This highlights how the use of pictures with an expressive illocutionary purpose can also be tied to an *arousalist* perspective of pictorial expression (as discussed in Sect. 2.4). Here, it is crucial to note the role of reflexive intention in determining the distinction between the performance of expressive acts and the mental states aroused in the spectator, which could be described as perlocutionary effects. If the use of the photograph is intentionally related to its capacity to evoke certain feelings in the spectator, and there is a reflexive intention that this intention be recognized by the spectator, we could classify this use as the performance of an expressive illocutionary act. This case can be exemplified by instances where the value of the photograph is manifest in its therapeutic capacity to evoke a certain state of mind in the viewer, which is recognized as a positive and beneficial state by both user and viewer (or part of its artistic value). But if the psychological state aroused in the viewer is not part of the user's communicative or reflexive intention, nor part of the purpose of the social practice, it would be appropriate to understand this psychological state as part of the *perlocutionary effect* of the photograph's use, rather than the performance of an expressive illocutionary act.

Furthermore, it is possible for a picture not to be expressive but still evoke feelings and mental states in the viewer. For example, a bland and uninteresting photograph might, for some strange reason, evoke anger in the viewer. In this case, we would say that it does not perform an expressive act but has a perlocutionary effect. We can also find this distinction in pornographic photographs. An image may be capable of sexually arousing the viewer even if, from a pictorial and aesthetical standpoint, it does not qualify as an expressive picture.

Thus, there would be a fundamental distinction determined by reflexive intention that allows us to differentiate the performance of *expressive acts* (whether from a personalist, arousalist or expression-look perspective) from the evocation of mental states as mere perlocutionary effects. This distinction enables us to separate the expressiveness of the picture (in Novitz's perspective), the performance of expressive illocutionary acts, and the perlocutionary uses of photographs (which would involve mental states aroused in the spectator, even if the picture itself is not expressive).

The significance of expressive acts in photography also lies in their role in facilitating other types of speech acts. For instance, an advertising photograph is more likely to succeed in its directive act of prompting viewers to purchase a product if it effectively performs a certain expressive act, imbuing the object with a particular mental state (whether overtly or covertly—in a Gricean or non-Gricean way). It could be argued that the directive act is a secondary act, contingent on the successful performance of the expressive act. An image that elicits feelings of disgust towards a depicted object would probably fail to persuade viewers to purchase that item. (We will later classify this case as an *illocutionary content condition* for performing the act (Sect. 4.3.2)).

3.5.5 Declaratory Photographic Acts

Declarations are speech acts in which the successful performance of the act aligns the world with the locutionary act, often occurring in contexts where the act determines the social or institutional status of someone or something. These acts have a dual direction of fit, as their successful performance brings the world into alignment with what has been stated. Thus, the declaratory act, in some specific sense (which we will need to later specify), would be *self-fulfilling*. A typical preparatory condition for these acts is that the individual performing them must hold the appropriate authority.

Let us consider some examples:

- "I declare the defendant guilty"—uttered by the judge during a legal proceeding.
- "I now pronounce you husband and wife"—uttered by a priest or minister in a wedding ceremony.
- "You are under arrest"—stated by a police officer.
- "This meeting is adjourned"—spoken by the chairperson of a meeting.
- "The game is over"—declared by a referee in a sports game.

In these cases, the utterance made by the person with the correct institutional authority would have the power to align the world with what was stated. By saying "I declare the defendant guilty," the judge changes the defendant's status, who, from the perspective of the legal process, becomes "guilty," thereby bringing the social world into alignment with what was stated. The same applies to the remaining examples.

Could photographs have the power to alter or determine the social status of something or someone, thereby performing a *declaratory photographic illocutionary act*? My answer to this question is affirmative, and I believe that the use of photographs for performing declaratory acts is more common than we could initially suppose. However, if we accept that photographs can be used to perform declarations, a second question becomes important: does the photograph have the necessary authority to perform declaratory acts?

We can argue in favor of the thesis that photographs can perform declaratory acts by revisiting Rae Langton's (2009) viewpoint on pornography (which we examined in Sect. 3.2.2). Langton uses Austin's taxonomy and classifies one of the speech acts performed by pornography as the *verdictive* speech act of *subordinating* women. Although Austin's (1962) verdictive acts can be related to Searle's (1979) assertive acts, the role pornography plays in relation to social hierarchies, as emphasized by Langton, allows us to analyze the case of pornography as an instance of what Searle calls *declarations*. The illocutionary force of pornography is not to assert that women are subordinate to men (in the sense that such an assertion could simply be true or false), but rather that pornography is endowed with the power to institute this form of subordination as a social fact, thereby bringing the world into alignment with what is depicted. This dual direction of fit in its social dimension (and not merely as a perlocutionary/causal effect, as we shall see below) is the distinguishing feature of declaratory acts. Metaphorically speaking, it is as if the use of pornography performs the declaratory act of saying, "I declare women to be subordinate to men."[18]

[18] This declaration does not need to be understood in the *imperative* sense, as Judith Butler (1997: 65) suggests. There are different ways to perform declaratory acts without using sentences that are grammatically imperative.

A critical aspect of Langton's analysis is the authoritative nature of the act of subordination, which necessitates a certain level of authority as a prerequisite. This is a fundamental point, as declaratory acts are typically authoritative. Without the necessary authority, an utterance such as "I declare so-and-so" will not have the power to establish a social fact. Langton argues that the authority of pornography is an empirical matter, due to the role attributed to it in instructing its consumers about "which moves in the sexual game are legitimate" (2009: 45). We could adapt this idea to Searle's taxonomy by suggesting that pornographic photographs, once they are granted *directive* illocutionary force (as they are regarded as *instructions* on how to behave in sexual contexts), can acquire the necessary authority to perform a declaratory act that ranks women as socially inferior. This occurs insofar as viewers engage with these images in ways that confer upon the images a certain degree of authority.[19] In simpler terms, we could say that pornography declares women to be socially inferior by legitimizing and promoting a certain type of behavior toward them. Thus, there would be a *self-fulfilling* aspect to the use of pornographic photographs, in the sense that the portrayal of women as subordinate to men becomes a social reality for women (with a double direction of fit).

This allows us to observe that, in this case, the logical role played by the photograph in the speech act is not merely propositional but pertains to the illocutionary force itself. It is the way pornography portrays women, together with the use to which the picture is put, that grants the photograph its declaratory illocutionary force.

If Langton's analyses are correct, and I believe they are, photography has the power to perform declarations, even if the necessary authority for this type of act is not directly related to an institution typically associated with authority. Photographs perform declaratory acts when they are used with the capacity or in relation to the capacity to determine the social status of the people or objects depicted. The authority required to perform the act may depend on the role these pictures have for their viewers and users, in which they ascribe an authoritative role to the photographs.

The declaratory power of photographs can, I believe, also be correlated with what Dominic Lopes describes as "weaponized aesthetics" in his 2024 book—though not every instance of weaponized aesthetics may constitute a declaratory act. A paradigmatic case of "weaponized aesthetics" involves the use of aesthetic resources to sustain structures of domination and exclude social groups by reinforcing negative stereotypes. The stereotypes may contribute to social injustice in two ways:

> In the first place, they licence the behaviours that harm people as members of subordinated groups. (…). In addition, it normalizes both the acts and the resulting interlocked harms by making them appear natural, expected, rational, or justified. (Lopes 2024: 9).

It is a form of *symbolic violence* that operates by normalizing aesthetic standards in ways that institute unjust social hierarchies and legitimize harmful patterns of

[19] The weakening of this authority would be a way to curb the harms of pornography against women, as Langton (2009) argues.

behaviors. From this perspective, Langton's account of pornography could be understood as an instance of "weaponized aesthetics." The declaratory dimension of stereotypes as weaponized aesthetics lies in the dual direction of fit through which the image acquires the power to constitute as social reality that which it depicts. Such stereotypes can have the power to *declare* these individuals to be socially inferior. This can certainly be applied to the case of photography, which may function as a form of weaponized aesthetics.

However, photographic declaratory acts are not necessarily negative. I think it is reasonable to claim that many photographs used in political propaganda aim to perform declaratory acts, determining the social status of what is depicted through the manner in which it is portrayed. A successful photographic campaign that portrays a certain politician in a particular way can positively (or negatively) determine that person's social status. As we shall see next, some advertising and many common uses of photographs on social media today can also be analyzed as related to declaratory acts.

It is crucial to highlight the extent to which the declaratory act differs from assertive ones. Let us explore the example of a photograph used as part of a political campaign to establish the social status of a candidate. An assertive act has a direction of fit picture-to-world, which makes the determination of truth or falsity relevant (the correspondence between the picture and the world). A photograph used to perform a declaratory act, by portraying a politician as confident and friendly, seems to assert this about the individual: "the candidate is confident and friendly"—potentially being true or false. However, from a declaratory perspective, the central element is not whether the individual is actually as depicted, but whether the image acquires the necessary power to establish this as a *social fact* by convincing people that the politician is as portrayed. If the political photographic campaign is successful in convincing a large portion of the public, the politician's *social status* becomes that of someone confident and friendly, even if this is factually false. In other words, this becomes the individual's social reality because it was determined by the successful photographic campaign, even if, from an assertive point of view, it is false. In this case, the photograph had the power to determine the individual's social status.

The use of photographs in advertising may also be interpreted as having a declaratory aspect when it aims to establish that certain objects or places are associated with a particular social status. A photograph has the power to determine, for example, that a particular product is a symbol of status for wealthy and successful individuals. It is as if, through the use of the photograph, the following performative act is carried out: "I declare this object is a symbol of wealth and success." It is the manner in which the object is depicted, together with the social context of use, that introduces this declaratory illocutionary force. And if people accept that this is the case, it thereby becomes the case.

But this leads us to the second question: does the photograph have the necessary authority to perform declaratory acts? We have seen that Langton's response to this question would be affirmative. In the examples from the previous paragraphs (in the cases of negative stereotypes, political propaganda, and advertising), the necessary authority for photographic declarations could be explained as emanating from a collective of people. For instance, a photograph used in political propaganda that leads a significant portion of society to accept a particular view about the subject depicted can have a substantial impact on that subject's social status. If such a belief becomes widespread, it may bring about a change in the social status of the person portrayed, thereby performing felicitously the declaratory act. The authority for this declaratory act arises from the *collective* that forms public opinion. The same can be said for the case of negative stereotypes and advertising, in determining the social status of objects and places. *If people believe it to be so, it becomes so from the point of view of social reality.*

The take-home message is that the way people or objects are portrayed in photographs often can determine or change their social status. This social power of photography can be understood as the performance of declaratory acts. And the authority required to perform these acts can stem from the collective authority of people who form public opinion (the viewers of the photograph), since by accepting what the picture depicts, it becomes a social fact. In other words, the declaratory force of photography would be associated with its capacity to construct social reality and determine the social hierarchies of objects and people, by convincing people that this is a social fact—what Lopes describes as "making them appear natural, expected, rational, or justified" (2024: 9).

We can further explore the distinction between declaratory and assertive acts in relation to the use of photographs on social media. It is particularly evident in social media that individuals frequently create photographic narratives to portray themselves in a certain manner, not to describe their actual social status, but to elevate themselves to a desired status. These photographs do not depict how they *are*, but how they *wish* to be seen in relation to social hierarchies.[20] Successful performance of this photographic speech act can lead to the individual acquiring the status portrayed. If spectators accept that this is the social status of the user of the photograph, it becomes their status (even if what is depicted in the photograph was not true as an assertive act). There is thus a *Munchausen* aspect, as these photographs possess the social power to *bootstrap* someone into a social status they did not originally have, in a self-fulfilling manner.

But it is important to note the distinction between the self-fulfilling dimension of the illocutionary act and a self-fulfilling dimension that could be described as a perlocutionary effect. The elevation to the desired social status through the use of photographs could ultimately be realized by the fact that social media engagement can be monetized. A person who portrays themselves as rich and famous, even if they are not, when accepted by a large number of people as having that social status, may effectively attain that status—not only by becoming recognized as famous and rich, but also by the ability to convert that fame into financial gain. This is the case described by Langton (2009: 105–106) as the *causal* and *perlocutionary* sense of *self-fulfilling*, in which the social change facilitated by the photograph is not only within the realm of social reality but extends to the actual world. The same could be

[20] On social hierarchies, see Wietmarschen (2021).

said for a political campaign that portrays someone as confident and decisive. If this becomes their social reality (in the illocutionary sense of self-fulfilling), the person might become more confident and decisive, thereby making reality what was portrayed (in the perlocutionary and causal sense of self-fulfilling). However, these would be contingent consequences of the declaratory act, which must be distinguished from the illocutionary force of the act from the perspective of illocutionary logic.

In the case of pornography, the distinction between the self-fulfilling dimension of the illocutionary act and the self-fulfilling dimension of the perlocutionary effects is crucial. For the *uptake* of the illocutionary act that declares women as subordinated to men, it is sufficient that, through the legitimization of a behavior (pornography as instruction), women are *recognized* by men as subordinated to them. Pornography becomes self-fulfilling from the perlocutionary perspective when men *act* based on the recognition of this subordination, creating real subordination. In other words, women are not only portrayed in pornography as subordinated to men (as a locutionary act) and subsequently recognized as subordinated (with the uptake of the illocutionary act), but real actions in the world can generate real effects based on how they are socially recognized as subordinated (this being the perlocutionary dimension).

Certainly, there are also cases in which the performance of declaratory acts is tied to individual authority and institutional recognition. A powerful person can determine the social status of something through the use of photographs. This is a frequent case in the artworld. In the realm of artistic photography, the authoritative dimension of speech acts is often related to the photographer's fame and public and institutional recognition. This enables them to perform acts that could be described as declaratory. An acclaimed photographer, by the mere act of portraying a particular object or person in a certain way, often has the power to elevate the social status of that object or person (or lower the status, in the case of a visual critique of what is depicted). There is also the sense in which a photograph has the power to determine what is depicted as art, performing the declaratory act that transmutes the social status of the object from ordinary to artwork (as if saying, "I declare this object to be an artwork").[21]

3.6 Maxims of Photographic Communication

Among the various ways we explored in Sect. 3.2 in which photographs can be related to speech acts, one significant approach is through the performance of *indirect acts*. In Sect. 3.3, we saw that, according to John Searle (1979), the performance of indirect speech acts presupposes, in addition to the theory of speech acts,

[21] The declaratory dimension of assigning the status function of artwork is explored by Catherine Abell (2012) as part of an institutional definition of art.

an understanding of the background knowledge necessary for inferring the indirect act, which includes general principles of cooperative conversation (as outlined by Paul Grice (1957)—which we analyzed in Sect. 1.3). My objective in this chapter is to question whether the principles and maxims described by Paul Grice have a photographic counterpart and are at work in photographic communication. The answer I will provide to this question is affirmative and will allow us to draw parallels between Searle's notion of indirect speech acts and Grice's notion of implicature, enabling us to develop the idea of *photographic implicatures*.

As seen in Sect. 1.3, Grice posited that communication is a rational cooperative exercise, governed by principles and maxims, based on "a common purpose or set of purposes, or at least a mutually accepted direction" (1989: 26). Grice distinguishes four categories of maxims and submaxims that govern communication (1989: 26–27)[22]:

Maxim of Quality: "Try to make your contribution one that is true."

1. Do not say what you believe to be false.
2. Do not say that for which you lack adequate evidence.

Maxim of Quantity:

1. Make your contribution as informative as is required (for the current purposes of the exchange).
2. Do not make your contribution more informative than is required.

Maxim of Relation: "Be relevant."

Maxim of Manner:

1. Avoid obscurity of expression.
2. Avoid ambiguity.
3. Be brief (avoid unnecessary prolixity).
4. Be orderly.

The adherence to these maxims forms the foundation of Grice's *Cooperative Principle*, which asserts that participants in a conversation should make contributions that promote effective communication while respecting these maxims.

The presumption that these maxims are either followed or evidently violated becomes the basis for *conversational implicatures*. For example, when someone standing in front of an elevator declares, "the elevator is broken," we assume that this statement is truthful and grounded in evidence (in accordance with the Maxim of Quality), that it is relevant to the ongoing context (in line with the Maxim of Relation), and that it is expressed in a clear and unambiguous manner (adhering to the Maxim of Manner). From these assumptions, we deduce that our interlocutor is referring to the nearby elevator and infer that waiting for it would be futile. If later, we discover that she was referring to a different elevator located elsewhere, we

[22] Due to the importance of these maxims in this chapter, I will cite them again as discussed in Sect. 1.3.

would be taken aback, as this would contradict our presuppositions regarding the Maxims of Relation and Manner. In cases of clear maxim violation, such as someone stating "what a beautiful sunny day" under a cloudy, stormy sky, the obvious falsehood (violating the Maxim of Quality) allows us to infer that this statement is ironic rather than literal.

In the realm of photographic communication, can we discern communicative maxims analogous to Grice's principles for verbal communication? One way to support this is by recognizing that in the everyday use of photographs, numerous presuppositions and inferences are often made. Let us examine some typical scenarios involving the use of photographs to scrutinize these presuppositions and inferences.

Scenario 1: A colleague talks about her recent international trip for an academic event. During the conversation, she opens her phone's photo gallery and shows a picture of a group of people at a restaurant. In a typical conversation, even without explicit confirmation, we would immediately assume that the photograph is from the trip and features the event's attendees.

Scenario 2: Upon opening a newspaper, you see an article discussing the rise in violence over the past year in a specific urban area. The article features a photograph depicting a violent scene. Although the caption may not specify this, you would likely infer that the image depicts the area mentioned in the article and was taken during the period relevant to it.

Scenario 3: On a social media platform, a colleague shares a photo of a dish in a picturesque restaurant on a Saturday night. Although not stated, you assume that the photo is recent, likely taken on the same evening the picture was posted, and that the post is meant to convey her presence at the restaurant.

Scenario 4: An artist organizes a photographic exhibition focused on the Native peoples of the Amazon Rainforest. The exhibit features various uncaptioned photos of Native individuals in a rainforest setting, with dense foliage and prominent rivers. It is understood that all these images were captured in the Amazon region by the artist.

In all these scenarios, if the inferences turned out to be incorrect, it would be quite surprising, potentially indicating bad faith on the part of the picture's user, possibly even leading to legal repercussions in some cases. For instance, in Scenario 1, if the photo was not from the mentioned trip or, more bizarrely, depicting someone resembling the colleague in a different context, it would be considered inappropriate to use the photograph in such a conversational setting. In Scenario 2, if the newspaper's photograph didn't depict the article's described location or its inhabitants, it would be perceived as an editorial error. In Scenario 3, discovering that the photo does not depict a place visited by the colleague or a dish not eaten by them would suggest deception. Similarly, if the Native peoples of the exhibition were actually shot with actors in a São Paulo studio with digital additions of vegetation and rivers, it would be a significant misrepresentation.

It's not crucial for our current discussion to meticulously categorize and dissect instances where our inferences based on photographs turn out to be false. The key observation is that such inferences are commonplace in conversational contexts involving photographs. These inferences play a vital role in determining what the picture depicts and denotes (from the perspective of *near-side* pictorial pragmatics), the communicative intention of the user, and its role in communication (which speech act is performed—from the perspective of *far-side* pictorial pragmatics). The essential point here is our reliance on inferences, which indicates that communication through photographs operates under its own set of maxims, akin to those outlined by Grice for verbal communication. In Grice's framework, these inferences facilitated by photographs could be termed *photographic implicatures*, and the overarching principles guiding these inferences could be referred to as *Maxims of Photographic Communication*. These maxims, while parallel to Grice's maxims in verbal communication, can be adapted to the specificities of visual communication through photographs.

But what maxims govern our communication through the use of photographs?

Certainly, in various photographic contexts, we observe the presence of a principle similar to Grice's Maxim of Quality. The photographic version would be something akin to:

- Photographic Maxim of Quality: the photograph should not provide false visual information about the depicted objects or events.

In a *quality context*, participants assume that a photograph should convey truthful information about the depicted objects or events. This notion is applied to pictures by Daisy Dixon (2018). As noted by Dixon, not all uses of pictures fall into quality contexts. In *non-quality contexts*, pictures are used without the presupposition that it provides true and accurate information about the world. An example can be found in aesthetic and artistic practices, where the primary purpose of using pictures is often expressive and may not be assertive. In these cases, there may be no presupposition that the photographs are truthful or that they accurately depict the world.

A philosophically intricate and highly pertinent issue lies in the problem of ascertaining the nature of truth and falsehood. There exist various philosophical theories of truth, and exploring these theories would extend beyond the scope of this book. For our current purposes, we can circumvent the excessive complexity of this problem by focusing on analyzing how, in everyday contexts, we determine the truth or falsehood of *photographs*—since our aim is to comprehend how we communicate through photographs. Therefore, we are not concerned with the metaphysical problem of truth in general or truth in photography (whether photographs themselves can be true or false); instead, our focus is on how we *use* photographs in situations where assessing the truth or falsehood of the image becomes pertinent. (This may be understood as part of the pragmatic turn, developed here as an extension of Dominic Lopes' "social turn" (2016), discussed in Sect. 1.5).

In everyday contexts, it is not common to evaluate the truth of a photograph based on the requirement of achieving a *complete correspondence* between the visual information presented in the image and the visual characteristics of the depicted object (or how the object appears from a specific vantage point). Such an approach would likely be impractical, as it would risk categorizing almost all photographs as false. Considering that objects are three-dimensional, the projection of an object's image onto a two-dimensional surface typically does not replicate the object's actual appearance, rendering complete correspondence unattainable.

The assessment of truth or falsehood is typically made in relation to the *purpose* of the photograph within its *context* (which, in Chap. 4, I will refer to as the *illocutionary purpose* of photographs, determined by the function of the practice, established by the collective intentionality of the participants). When we employ a photograph to make an assertive speech act (expressing a commitment to the truth of what is depicted), the determination of what visual information will be judged as true or false depends on which aspects of the pictorial content are relevant for communication in that context and moment. It does not encompass all instantiated and identifiable visual properties within the picture. Additionally, there are contexts that are more lenient, allowing for a greater degree of deviation between the visual information presented and the visual properties of the depicted objects. Conversely, there are contexts that impose stricter requirements, demanding greater precision and control over the visual information provided for it to be deemed true.

A quality context can be defined as follows:

- Quality Context is one in which users presuppose that the visual information present in the pictorial content, *relevant to communication in the context*, is true and accurately describes relevant facts.

However, in terms of reference theories, we would say that photographs denote not only through their descriptive visual properties but primarily through the causal route that connects them to the depicted objects. When we say that a photograph is a "photograph *of* an object X," we generally mean that object X was the causal referent of the photograph (that it was in front of the camera constituting the pro-photographic scene). Hence, even if the picture faithfully depicts the object (in a descriptive sense), if the recognized depicted object is not the one that served as the causal referent, we would typically assert that the photograph is false. A photograph that visually accurately describes President Luiz Inácio Lula da Silva but was made using a look-alike would (depending on the context) be deemed a false photograph of the president. Thus, there is a difference between a photograph being *accurate* and being *genuine*. Genuineness relates to the correct causal genesis of the picture, while accuracy pertains to the correspondence between the visual properties of the picture and the world.

Therefore, truth or falsehood in a photographic context is not determined only in relation to the visual properties of the pictorial content—in a descriptive manner. In everyday photographic contexts, we typically assume that the object visually recognized in the pictorial content is indeed its causal referent. This allows us to propose an extension of the maxim of quality, in causal terms:

- Photographic Maxim of Causal Quality: the objects or events considered in the context as depicted by the photograph must be the causal referents of the picture.

It is essential to remember that there are contexts in which the photograph can be taken using substitutes (such as maquettes and models), so that the photographic maxim in the descriptive sense is maintained while the causal one is abandoned. This case is common in advertising photography. In advertising photography, it is accepted that an image can be created using a mock-up, which belongs to a different ontological category, as long as it accurately describes the object.

Additionally, photographs can have more or less information. Out-of-focus pictures, for example, provide less visual information about the scene than a completely focused image. Photographs can contain an immense amount of visual information, depending on the resolution of the system used and the visual characteristics of the scene. For successful communication, it is important to also have a *Maxim of Quantity in operation.* In photographic terms, we could formulate this maxim as follows:

- Photographic Maxim of Quantity: the photograph proposed as a candidate for attention in the context must be as informative as necessary for communication.

The anecdote recounted by Nelson Goodman (previously addressed in Sect. 2.3) illustrates the importance of the Maxim of Quantity. Goodman considers a case in which he tells someone that he has "a picture of a certain black horse." However, when showing the image, the horse "come out a light speck in the distance" (1976: 59–60). In this case, the photographic Maxim of Causal Quality would have been respected because the horse is the causal referent of the image, but the amount of visual information present in the pictorial content would be less than adequate for communication. The point at issue is that, in our interlocutory use of photographic images, we presuppose that the image should have an appropriate amount of visual information for the context. The interlocutor's feeling of being deceived arises from the fact that they presuppose that the photograph suitable for the context should be one in which it would be, at the very least, possible to identify the depicted horse and its visual properties.

Failure due to an excess of information would also constitute a violation of the Maxim of Quantity, resulting in ineffective communication. For instance, if I were to claim that I possess a photograph of my grandfather and then present a photograph depicting a large crowd with thousands of people, among whom my grandfather is indeed a part, analogous to Goodman's case, my interlocutor can hardly accuse me of lying. Nevertheless, the sheer volume of information in the image would not be beneficial to effective communication unless further specification is provided—such as using a gesture to point out my grandfather within the crowd.

Another maxim that certainly plays a role in communication through photographs is the Maxim of Relation, which dictates that contributions in a conversation must be relevant to the ongoing communication context. In the realm of photography, a corresponding maxim could be formulated as follows:

- Photographic Maxim of Relation: The photograph presented as a candidate for attention in the context must contribute in a relevant and meaningful way to the communication.

Many of the scenarios discussed earlier in this chapter involved implicatures based on the Maxim of Relation. When a particular topic is under discussion, we presuppose that photographs presented for consideration by interlocutors are related to the context in a meaningful way (as seen in scenarios 1 and 2).

More generally, we also assume that an image proposed for attention in a conversational context aligns with the sequence of ideas being discussed and is presented in a clear manner. Thus, in a conversation that covers various topics sequentially, we presuppose that the image relates to the currently discussed (or recently discussed) theme. Consequently, we can posit a Maxim of Manner in photography:

- Photographic Maxim of Manner: The communicative use of the photograph must be clear and coherent.

Regarding scenario 1, if our interlocutor were to show us a photograph that was not from the mentioned trip, it would constitute a violation of the Maxims of Relation and Manner. The photograph would not be meaningfully related to the discussed subject, and the communicative contribution of the photograph would lack clarity and coherence. The discovery that the image is not causally related to the discussed subject would have different consequences depending on whether the violation of the maxims is intentional or unintentional. If the breach of the Maxim of Relation were unintentional, one might surmise that the interlocutor made a mistake. However, if the violation is intentional (with the proponent knowingly breaching such maxims), the potential consequences depend on whether this breach is overt or covert. If the proponent clearly shows that they are intentionally violating the maxims, the communicative role of the image for the discussed subject may be indirect. For example, it could be used for irony or metaphorical expression. (An old photograph from a different trip can be used to express that the current trip was as interesting and enjoyable as the past one.) However, if the violation is intentional but covert, this would likely be a case in which the proponent intends to deceive. In this scenario, the interlocutor assumes that the Maxims are being respected, while the proponent knowingly disregards them. The use of the photograph aims to cause the interlocutor to believe that it accurately represents a certain scene, and such a belief would be false. In such a case, we could conclude that the photograph also violates the Maxim of Quality because it would not be truthful in its intended illocutionary assertive purpose.

In the context of photography, the Maxim of Manner can also be considered in relation to the style and the aesthetics of the image. This maxim, when viewed from this perspective, suggests that effective communication also demands that the picture's style and aesthetics be suitable for the communicative context (a concept parallel to Searle's dimension of variation in the Differences in the Execution Style of the illocutionary act—which we will examine in Sect. 4.2). For example, it would not be appropriate to use photographs with a horror-movie film noir aesthetic in a classroom setting with children. Similarly, images with aesthetic elements typical of Pop Art would not be suitable for communicative use in medical diagnostic contexts. A second version of the Maxim of Manner in photography could be formulated as follows:

- Photographic Maxim of *Stylistic* Manner: The communicative use of the photograph should be executed in a manner that aligns with the aesthetics and style of the context.

Intentionally and obviously deviating from the aesthetic version of the Maxim of *Stylistic* Manner has the potential to introduce irony or humor into the use of the image. For example, depicting a horrifying scene using the typical colors of a children's drawing would be an example of such an evident departure, adding an ironic or tragicomic dimension to the picture's use (common in Pop Art).

The analyses conducted in this chapter lead us to conclude that Paul Grice's communicative maxims have their photographic counterparts and are indeed present in communication through photographs. We can schematically summarize the photographic versions of these maxims as follows:

> 1(a). Photographic Maxim of Descriptive Quality: the photograph should not provide false visual information about the depicted objects or events.
> 1(b). Photographic Maxim of Causal Quality: the objects or events considered in the context as depicted by the photograph must be the causal referents of the picture.
> 2. Photographic Maxim of Quantity: the photograph presented as a candidate for attention in the context must be as informative as necessary for communication.
> 3. Photographic Maxim of Relation: the photograph presented as a candidate for attention in the context must contribute in a relevant and meaningful way to the communication.
> 4(a). Photographic Maxim of Manner: the communicative use of the photograph must be clear and coherent.
> 4(b). Photographic Maxim of Stylistic Manner: the communicative use of the photograph should be executed in a manner that aligns with the aesthetics and style of the context.

It is essential to recognize that these maxims serve as general communication guidelines in the use of photographs, but there are contexts in which they may be suspended. For instance, in a non-quality context, the Photographic Maxim of Quality will not apply. Furthermore, different contexts will establish varying standards for what constitutes an appropriate quantity of information, relevance, manner, orderliness, and suitable style and aesthetics in communication through photographs. Therefore, the determination of which maxims are in operation and the specific requirements associated with each maxim will vary from context to context. This contextual variability influences how communication through photographs occurs. This indicates that the discussion about the truth or falsity of photographs should not be framed as a metaphysical problem, but rather should begin with an understanding of which context-sensitive elements are at play and are decisive in assigning truth or falsity to pictures within the context.

References

Abell, C. 2012. Art: What it is and why it matters. *Philosophy and Phenomenological Research* 85 (3): 671–691.

Atencia-Linares, P. 2012. Fiction, non-fiction, and deceptive photographic representation. In *The media of photography. Special issue of Journal of Aesthetics and Art Criticism*, ed. D. Costello and D. M. Lopes, vol. 70. Winter 2012.

Austin, J. L. 1962. *How to do things with words*. London: Oxford University Press.
Barthes, R. 1981. *Camera Lucida: Reflections on photography*. Trans. R. Howard. New York: Hill and Wang.
Butler, J. 1997. *Excitable speech: A politics of the performative*. New York and London: Routledge.
Da Silva, G. 2018. What does a presentist see when she looks at photographs of dead relatives? Laocoonte. *Revista de Estética y Teoría de las Artes* 5:97–116.
Da Silva, G. 2023. Quatro formas de silenciamento: Um estudo a partir da crítica de Rae Langton à pornografia. *Philosophus* 28 (2): 123–145. UFG.
Danto, A. C. 1981. *The transfiguration of the commonplace: A philosophy of art*. Cambridge, MA: Harvard University Press.
Dixon, D. 2018. Alterpieces: Artworks as shifting speech acts. PhD diss., University of Cambridge.
Fodor, J. 2012. Déjà vu: All over again: How Danto's aesthetics recapitulates the philosophy of mind. In *Danto and his critics*, ed. Rollins, 2nd ed. Oxford: Blackwell Publishing.
Goodman, N. 1976. *Languages of art: An approach to a theory of symbols*. 2nd ed. Indianapolis: Hackett.
Grice, P. 1957. Meaning. *The Philosophical Review* 66 (3): 377–388.
Grice, P. 1989. *Studies in the way of words*. Cambridge: Harvard University Press.
Hopkins, R. 2015. The real challenge to photography (as communicative representational art). *Journal of the American Philosophical Association* 1 (2): 329–348.
Kjørup, S. 1978. Pictorial speech acts. *The Philosophy of Nelson Goodman* 12 (1): 55–71.
Korta, K., and J. Perry. 2011. *Critical pragmatics: An inquiry into reference and communication*. Cambridge: Cambridge University Press.
Korta, K., and J. Perry. 2020. Pragmatics. In *The Stanford encyclopedia of philosophy (Spring 2020 Edition)*, ed. E. N. Zalta. Available at: https://plato.stanford.edu/archives/spr2020/entries/pragmatics/.
Kulvicki, J. 2020. *Modeling the Meanings of Pictures: Depiction and the Philosophy of Language*. Oxford: Oxford University Press.
Langton, R. 2009. *Sexual solipsism: Philosophical essays on pornography and objectification*. Oxford: Oxford University Press.
Lopes, D. 2004. Directive pictures. *Journal of Aesthetics and Art Criticism* 62 (2):
Lopes, D. 2005. *Sight and sensibility: Evaluating pictures*. Oxford: Clarendon Press.
Lopes, D. 2016. *Four arts of photography*. Oxford: Wiley-Blackwell.
Lopes, D. 2024. *Aesthetic injustice*. Oxford: Oxford University Press.
MacKinnon, C. 1987. *Feminism Unmodified*. Cambridge, Mass.: Harvard University Press.
Novitz, D. 1977. *Pictures and their use in communication: A philosophical essay*. Springer.
Searle, J. 1969. *Speech acts: An essay in the philosophy of language*. Cambridge: Cambridge University Press.
Searle, J. 1979. *Expression and meaning: Studies in the theory of speech acts*. Cambridge: Cambridge University Press.
Searle, J. 2010. *Making the social world: The structure of human civilization*. Oxford: Oxford University Press.
Sperber, D., and D. Wilson. 1995. *Relevance: Communication and cognition*. Oxford: Blackwell Publishers Ltd.
Sperber, D., and D. Wilson. 2004. Relevance theory. In *The handbook of pragmatics*. ed. L. R. Horn and G. Ward, 607–632. Oxford: Blackwell.
Strawson, P. F. 1964. Intention and convention in speech acts. *The Philosophical Review* 73 (4): 439–460.
Wietmarschen van, H. 2021. What is social hierarchy? *Noûs* 56 (4): 920–939.
Wittgenstein, L. 1953. Philosophical investigations. Trans. G.E.M. Anscombe. Oxford: Blackwell.
Wollheim, R. 1980. *Art and its objects*. 2nd ed. Cambridge: Cambridge University Press.

Chapter 4
The Ontological Foundations of the Photographic Social Practices

Abstract This chapter lays the foundation for an ontology of photographic social practices, presenting a philosophical framework that examines how photographs function as vehicles for communication through speech acts. It builds on Dominic Lopes's "social turn" by incorporating ideas from John Searle's social ontology, emphasizing that photographic practices are rooted in collective intentionality. These practices have a perceived function that defines the illocutionary purpose of photographs within the practice. The chapter analyzes how the felicity conditions for the uptake of speech acts, understood as the illocutionary purposes of the practices, can function as constitutive and regulative rules. A typology of the conditions that serve as rules will be proposed. We will examine how adherence to the rules of the practice enables the attribution of a status function to a photograph related to the practice and, consequently, the assignment of deontic and epistemic powers. The chapter also distinguishes between conventional and non-conventional contexts. In conventional contexts, adherence to constitutive rules allows for non-intentional communication, where the mere use of a photograph implies a communicative intention linked to the practice's perceived function. In contrast, non-conventional contexts rely on Gricean reflexive intention to determine the illocutionary force of photographic communication.

Keywords Illocutionary purpose · Collective intentionality · Constitutive rules · Content conditions

The goal of this chapter is to establish the foundations for an ontology of photographic social practices, allowing us to better understand how photographs are used to communicate through photographic speech acts. This will not be an empirical study of existing photographic social practices (how they are actually structured and defined), but rather a proposal of theses and concepts that can be used to think about the structure of photographic practices and how it impacts communication. This is

G. Ghisoni da Silva, *The Philosophical Foundations of Photography as a Means of Communication*, Synthese Library 526,
https://doi.org/10.1007/978-3-032-02649-1_4

an eminently philosophical task.[1] This theory aims to extend and deepen the "social turn" proposed by Dominic Lopes (2016). I will further develop the notion of photographic social practice through some of John Searle's theses on *social ontology* (1995, 2010).[2] The importance of Searle's social ontology lies in its provision of a framework for understanding how photographic social practices are structured and how they can relate to the speech acts that photographs are used to perform within these practices.

But before we turn to the main topic, an important caveat is in order. Although I will attribute a certain structure to photographic social practices, I do not claim that *all* possible photographic practices necessarily conform to the structure I will outline. I suspend judgment concerning the universality of the framework I am about to propose. Thus, I acknowledge the possibility that some practices may lack the structure I ascribe to them. What I offer here is not a general theory of all photographic social practices, but rather a conceptual framework aimed at illuminating those practices that are embedded in our everyday lives.

The central theses I propose, which constitute the core of this book, are as follows. A photographic practice is a *social fact*, since it "contains a collective intentionality of two or more human [...] agents" (Searle 2010: 156).[3] People with similar goals and intentions may form a collectivity related to the use of photographs, recognizing themselves as performing actions that share common goals and intentions. They recognize themselves as acting in a social context with a *perceived function* that determines the goal towards which their intentions converge.[4] Thus, participants in a photographic social practice share a *collective intentionality* that determines the *perceived function* for which the photographic practice exists. Different practices will have different perceived functions. For example, photojournalism, advertising photography, forensic photography, political propaganda photography, and the use of photographs in artistic and scientific contexts are photographic social

[1] Thus, I admit that the descriptions I offer of the function of certain practices and the purpose of photographs within these practices will be superficial and imprecise. An accurate description of existing practices and the purpose for which photography is intended within these practices would require an empirical study, which lies beyond the scope of this book.

[2] The ideas presented in this initial section of Chap. 4 are the development of the photographic version of the key concepts found in Searle (1995, 2010). These concepts appear in italics when introduced in the text. As mentioned in a footnote in the introduction to this book, the idea of connecting photographic social practices with Searle's social ontology was inspired by my reading of Catharine Abell's insightful article from 2012. Abell uses John Searle's account of institutions and institutional facts to provide a reductive institutional definition of art. Here, I set aside the discussion from the problem of defining art.

[3] For a critique of Searle's notion of collective intentionality, see the collection edited by Savas L. Tsohatzidis, dedicated to the themes of *Intentional Acts and Institutional Facts* (2007).

[4] As emphasized by Catharine Abell, what is essential is not the actually performed function, "but that its participants believed it to do so" (2012: 12). Thus, a practice "that never actually performs any valuable function may nonetheless come into existence so long as its participants believe it to perform such a function or functions" (idem). This is why the *perceived* function attributed to a photographic social practice by its participants is crucial, rather than its *actual* function.

practices with different perceived functions, each attributing to photography a different *purpose*.

In Lopes's analysis (2016), the purposes of photography in relation to the perceived function of social practices are framed by the distinction between epistemic and non-epistemic, and artistic and non-artistic purposes. In the theory I outline here, the purposes of photography will be understood as the specification of the type of speech act (or speech acts) that photographs primarily perform in relation to the perceived function of the practice. Thus, I will call the purpose of photography in relation to the perceived function, drawing on Searle, the *photographic illocutionary purpose*.[5] The illocutionary purpose of photographs fixes the primary *communicative intention* attributed to the photographs in that practice.

Photographic practices that have historically been consolidated generally include the creation of *institutions* of different types and require different *rules* for participation. Thus, photographic social practices may contain institutions of various types, with participants performing different roles. Engagement in photographic practices occurs as follows. Photographs are produced and used according to the rules of the practice, with the intention that they be understood and appreciated in relation to the purpose of the photographs as determined by the perceived function of the practice. Furthermore, the participants in the practice engage in accordance with the rules of the practice on the condition that other participants do so too.

Conventional practices establish *constitutive rules* to ensure that the illocutionary purpose of pictures is achieved, thereby contributing to the fulfillment of the perceived function. These rules must be followed for the photograph to be accepted within the practice. For example, in the context of forensic photography, it is imperative that pictures adhere to rigorous production and editing rules and be produced by a person with the correct authority in relation to certain extralinguistic institutions. Photographs that do not conform to these rules will not be accepted as forensic photographs for the purposes for which they are intended within the institutions related to the practice. A practice may also have *regulative rules*, which determine elements that are commonly expected and accepted in the practice, without noncompliance preventing the use of the photograph in the context.[6]

The function of the practice often plays a teleological role in the production and development of photographic technologies. If the perceived function determines a certain purpose of photography that requires it to be capable of producing pictures of a certain type (for example, with a certain degree of epistemic reliability), the technology used in the practice may be developed and improved with the aim of achieving the end sought. Significant technological achievements are those that enhance the fulfillment of the purpose that photography serves within the practice with respect to the practice's perceived function.

---

[5] As we will see in the following sections, practices may have more than one illocutionary purpose. This would determine a space of possibilities related to the combination of the five types of acts in Searle's taxonomy (1979).

[6] On constitutive rules, see Searle (1995: 28–29, 2010: 96–97). On regulative rules, see Searle (1995: 29, 2010: 97). We will explore this distinction thoroughly in Sect. 4.2.

The teleological aspect of the function of the practice may also foster a theory of value. Photographs will be more highly valued within the practice if they tend "to ameliorate the institutionalised pursuit of the functions to the perceived performance of which [...] institutions owe their existence" (Abell, 2012).[7] That is, they will be more highly valued insofar as they make significant contributions to the fulfillment of the practice's perceived function.

The fulfillment of constitutive rules enables the acceptance of a photograph within the practice, which confers upon it a certain *status function* (Searle 2010: 99–100). For example, once the conditions are met for forensic photography, the photograph can be recognized as a *forensic photograph* (receiving that status function) and may serve as *evidence* in a crime trial. Similarly, if the conditions are met for participation in an artistic photographic practice, the photograph can be assigned the status function of serving as an *artwork* within the institutions that are part of the practice.[8] It is also possible that, in some practices, the assignment of a status function occurs through a *status function declaration*—such as a declaration issued by an individual with the appropriate institutional authority to perform the act (as may be the case in certain artistic contexts involving the act of declaring a photograph to be a work of art).[9] In the case of status assignment through a status function declaration, the role of constitutive rules is diminished—potentially leaving only the constitutive rule that a status function may be assigned by someone with the appropriate authority within the practice. My position is that a status function may be assigned in different ways depending on the context. One can distinguish between *conventional* and *non-conventional* contexts—understood here as ends of a *continuum* with varying degrees. Conventional contexts tend to prioritize the assignment of functions through the satisfaction of explicit constitutive rules, whereas non-conventional contexts may rely more heavily on status function declarations. Since my focus is on developing an ontology of photographic social practices, which are predominantly conventional in nature, I will concentrate on the assignment of status functions through the fulfillment of constitutive rules.

---

[7] In this context, I adapt Catharine Abell's theory of artistic value to the broader domain of photographic social practices, extending it beyond the artistic sphere to include non-artistic social practices as well.

[8] It is not my intention here to propose a definition of the concept of *art*, as I believe that each artistic practice may have its own standards for what counts as art and may adopt different definitions of it. In this regard, my theory of photographic communication could employ Dominic Lopes's (2014) "buck-passing theory". According to his theory, the definition of art should be deferred to the practices and social contexts in which artworks are situated, produced, and evaluated.

[9] In his 2024 article, Rodrigo Gouvea argues that Searle has shifted his position between *The Construction of Social Reality* (1995) and *Making the Social World* (2010). According to Gouvea, in 1995 Searle maintained that the assignment of a status function depended on collective intentionality and collective recognition that something or someone satisfied the relevant constitutive rules. In 2010, however, Searle, as Gouvea contends, gave priority to a language-based account, in which the assignment of status function takes place through "status function declarations". While I consider Gouvea's interpretation both accurate and philosophically pertinent as an exegetical account of Searle's work, I do not think it is necessary to choose between the two models.

Constitutive rules will take the following form: "X counts as a photograph in practice P if...". A crucial part of our analysis in Chap. 4 will be filling in the missing element after the conditional. This will be achieved through understanding what types of conditions a practice can establish to ensure its illocutionary purpose. I will propose that the constitutive rules can be understood based on the *felicity conditions* for the performance of the illocutionary act which is the purpose of photographs in relation to the perceived function of the practice. Thus, the "X counts as a photograph in practice P if..." formula will be completed by understanding the conditions for the *felicitous* performance of the illocutionary purpose of the practice, which the practice may require as a constitutive rule for participation. These conditions may include *preparatory* conditions, *content* conditions, and *sincerity* conditions, with the *essential* condition being that the illocutionary purpose is intended (Searle 1979: 31). Thus, the formula will then have the following structure: "X counts as a photograph in practice P if X fulfills the essential condition, and/or preparatory condition, and/or sincerity condition, and/or content condition."[10] The fulfillment of the conditional allows the attribution of a function related to a collective imposition in the form "this photograph X counts as having this status and therefore this function as Y in this practice P."[11]

Once a status function has been attributed to a photograph, it may, in many cases, acquire a specific kind of *power* (or set of *powers*) within that practice. As we shall see in the following sections, in some instances, the power bestowed upon the photograph by status function attribution will clearly be *deontic* in Searle's sense, involving *rights*, *duties*, *obligations*, *permissions*, *authorizations*, and the like.[12] For example, a picture assigned the status function of being an *advertising photograph* will have certain obligations attached to this use. As seen in Sect. 3.5, in advertising photography, there is the performance of a *commissive* illocutionary act, wherein the purchase of the depicted object *obligates* the seller to provide the buyer with an object similar to the one in the photograph (except in cases where the disclaimer "image for illustrative purposes only" is included). This obligation is a *negative deontic power* related to the status function attributed to the photograph as advertising photography.

However, in other cases, it may be more appropriate to adopt a terminology better attuned to the specificities of photography, departing from Searle's original framework. This applies, for example, to *epistemic normative powers* (powers related to knowledge, justification, evidence, or belief formation) attributed to

[10] Understanding these conditions will require developing the photographic version of Searle's *dimensions of variation*, which are the dimensions according to which speech acts can vary (1979). An important part of Chap. 4 will involve understanding the *content conditions* in relation to the stages that constitute photographic technology. For this purpose, I will adopt the perspective of the New Theory in the philosophy of photography, which identifies four stages of photographic technologies.

[11] Here, I adapt to photography the ideas of Searle, as presented in (2007: 12–13).

[12] On deontic powers related to status functions, see Searle (2010: 8–9, 123).

photographs that hold particular status functions within assertive practices. These powers may not be best understood as strictly deontic, yet they may nonetheless presuppose deontic foundations. Consider, for instance, a photograph that has been assigned the status function of a scientific astronomical image within the context of an institutionalized scientific practice, having fulfilled the constitutive rules of that practice. By virtue of this assignment, the photograph may acquire the *epistemic power* to disconfirm a given hypothesis. This power does not stem merely from the photograph as such, but from its collective recognition, within that practice, as meeting the criteria necessary to be granted that power. An identical photograph, lacking such a status, would not possess the power to perform that epistemic role.

From this perspective, the factive experience afforded by certain uses of photographs would not be part of photography's self-conception (as Robert Hopkins argued in 2015, as discussed in Sect. 1.1). Instead, factivity is a *power* related to the attribution of a status function within practices that ensure the fit of the pictorial content with the world, grounded in certain constitutive rules of image production. (We will explore this idea more thoroughly in the following sections.)

Given that the powers arising from the attribution of a status function may be deontic but can also include powers that are not directly deontic, being better described as epistemic powers, I will classify them more broadly as "photographic powers" (or simply "powers," for short). By *photographic powers*, I mean the powers a photograph comes to possess within a given practice once a certain status function has been attributed to it.

The crucial point to emphasize is that a photograph can be visually identical to another in terms of its vehicle properties and pictorial content, but if the first is produced in accordance with the constitutive rules of the practice and the second is not, they can still differ in terms of *status functions*. The status function of a photograph as journalistic, artistic, forensic, scientific, etc., depends on the community collectively accepting that those photographs have the status of journalistic, artistic, forensic, scientific, etc. This demonstrates that photographic pictures are underdetermined in relation to their means of production. Two pictures may be visually identical yet produced by different photographers through different procedures and technologies. I will refer to this phenomenon as *photographic underdetermination*. Consequently, when the means of production are regulated by the constitutive rules of a photographic practice, two visually identical photographs can hold different *powers*—because one may adhere to the practice's conditions while the other does not. This allows us to further develop the idea of a *communicational underdetermination* of photography (which can be added to the *pictorial* and *denotative underdeterminations* discussed in Chap. 2). Due to *photographic underdetermination*, visually identical photographs can have different status functions and powers, performing different speech acts. Thus, the status function of a photograph, and its powers, are not limited to the picture as an object (its vehicle properties and content), but also includes the recognition that the image plays a role in relation to the function of the social practice through the fulfillment of its constitutive rules (or through status function declaration).

The consolidation of the practice occurs through the establishment of institutions and the explicit articulation of constitutive rules that aim to contribute to the achievement of the practice's function. Those rules must be followed for the photograph to be part of the practice. Therefore, if the photograph is used within a conventional practice, participants can assume that the rules have been followed and the illocutionary purpose intended. If the rules had not been followed, the photograph should not be accepted within the practice, nor would it have the status function granted to it. This allows us to adopt a *non-intentionalist* approach to photographic communication within conventional practices, since intending the illocutionary purpose is an *essential condition* for participation. Thus, the user does not need to explicitly manifest their communicative intention, as we can attribute a certain communicative intention to the use of that photograph from the mere presence of the picture within the practice. For example, a photojournalist does not need to overtly manifest their communicative intention to perform an assertive illocutionary act in accordance with the constitutive rules of photojournalism every time they use a photograph, as the mere presence of the photograph in a newspaper (or in some context related to journalism institutions, such as books and exhibitions) already permits the attribution of intentionality to the use of the picture in accordance with the illocutionary purpose and the constitutive rules of the practice. As a result, we can ascribe to the photograph itself, in a derivative manner, the performance of the speech act and sidestep the importance of a Gricean reflexive intention in communication. In other words, it is not necessary to *explicitly* manifest the communicative intention, because intending a specific illocutionary act is a prerequisite for participation in the practice (as an *essential condition*).

This understanding of photographic social practices will allow us to draw an important distinction between conventional and non-conventional contexts. This distinction aims to differentiate practices that possess a high degree of internal regulation through the adoption of various types of constitutive rules from those with a low degree of regulation. In conventional contexts, the illocutionary purpose of photographs and the conditions for their use are already predetermined (although they may evolve over time). This can also determine the deontic or epistemic powers the photograph will possess within the practice, once a particular status function has been attributed to it. Thus, communication is structured based on the presupposition that the purpose has been intended and the conditions fulfilled. In non-conventional contexts, none of this can be presupposed and communication depends on the overt manifestation of the communicative intention of the user and the exchange of additional information, which ensures the fulfillment of the conditions for the implementation of the intended act. Thus, communication in non-conventional contexts will depend on a Gricean reflexive intention.

Since the notions of completely conventional and completely non-conventional contexts are merely theoretical constructs, what we encounter in everyday practices is a *continuum* in which practices exhibit varying degrees of conventionality. Thus, even in a practice where there are constitutive rules and the determination of the illocutionary purpose of photographs, there can be room for the reflexive intention related to secondary or indirect speech acts.

The ideas to be developed in Chap. 4 may be summarized as follows. Photographic social practices that qualify as conventional contexts generally exhibit a characteristic structure:

- The collective intentionality of the participants establishes a perceived function that fixes the purpose of photographs within the practice. This purpose can be understood as the illocutionary purpose of photographs, insofar as it specifies the speech act(s) that photographs are primarily intended to perform within the practice.

The rules that structure photographic social practices and regulate communication within them can therefore be conceived in the following terms:

- Conventional contexts are governed by constitutive rules that correspond to the felicity conditions of the speech acts that establish the illocutionary purpose of the practice. The constitutive rules of the practice may thus be articulated in relation to the essential conditions, the content conditions, the preparatory conditions, and the sincerity conditions of the speech act(s) in question.

This framework enables us to analyze the conditions under which photographic speech acts are implemented according to the following categories:

- Essential condition: illocutionary purpose
- Content condition: (i) pictorial rules, (ii) production rules, and (iii) aesthetic rules
- Preparatory conditions: (i) authority requirements and (ii) institutional linkage
- Sincerity conditions: required psychological state

The aim of the following sections of Chap. 4 is to clarify this structure of social practices and photographic communication. With these elements in hand, we can then establish the philosophical foundations of a theory of photographic communication for conventional and non-conventional contexts.

4.1 Five Illocutionary Forces as Dimensions of Variation

As outlined in the previous section, photographic social practices are structured by a collective intentionality that establishes the perceived function for which the practice exists. This perceived function determines the purpose of photographs within the practice. Since the purpose of photographs can be understood as the speech act (or acts) primarily associated with the practice, we can regard this as the *photographic illocutionary purpose*.

At this point, the taxonomy of photographic acts proposed in Sect. 3.5 allows for a taxonomy of photographic social practices. According to Searle, the five illocutionary purposes are:

We tell people how things are (*Assertives*), we try to get them to do things (*Directives*), we commit ourselves to doing things (*Commissives*), we express our feelings and attitudes (*Expressives*), and we bring about changes in the world through our utterances (*Declarations*). (1979: viii)

Categorizing photographic speech acts into five types facilitates the classification of photographic practices into five corresponding general categories: *assertive*, *directive*, *commissive*, *expressive*, and *declaratory*.[13]

Let us consider some general descriptions of the purpose to which photography is used within certain photographic social practices[14]:

- The purpose of photojournalism is to visually capture and convey information about events and facts.
- The purpose of advertising photography is to create engaging and visually appealing images to attract and persuade consumers to buy products.
- The purpose of photography in a product catalog is to present the appearance of the product to be purchased by the buyer.
- The purpose of photography in art is to convey emotions, express creativity, and explore artistic concepts.
- The purpose of photography in political propaganda is to manipulate and influence public opinion.

Regarding the taxonomy of photographic speech acts, it can be said that the illocutionary purpose of photojournalism is primarily *assertive*, the illocutionary purpose of advertising photography is primarily *directive*, the illocutionary purpose of photographs in product catalogs is primarily *commissive*, the illocutionary purpose of photography in art is primarily *expressive*, and the illocutionary purpose of photography in political propaganda is primarily *declaratory*.[15]

The idea that a practice is primarily assertive, directive, commissive, expressive, or declaratory should be viewed *cum grano salis*. The same photograph can be used to perform various distinct speech acts. Thus, it is not logically necessary for a practice to be restricted to a single illocutionary act as its illocutionary purpose. A practice could have more than one act as its illocutionary purpose. There is also the possibility of performing indirect speech acts (as discussed in Sect. 3.4). Practices could combine different illocutionary purposes in different numbers. This defines the logical space of illocutionary possibilities, even though in the actual world, there may not be a social practice that occupies all these positions within this logical space.

The logical space of combining two illocutionary purposes would give us 10 types of practices:

---

[13] This is only a general taxonomy, intended to serve as a starting point for future work that may refine and enhance these categories. I do not intend for this taxonomy to be definitive or to provide a detailed understanding of all possible uses. This will require finer-grained concepts.

[14] As mentioned in a previous note, I acknowledge that the descriptions I offer of the function of certain practices and the purpose of photographs within these practices may be overly general and imprecise. A more accurate description would require empirical study of the practices, which is beyond the scope of this book.

[15] How these categories apply to these cases was explored in Sect. 3.5.

> Assertive and Directive, Assertive and Commissive, Assertive and Expressive, Assertive and Declaratory, Directive and Commissive, Directive and Expressive, Directive and Declaratory, Commissive and Expressive, Commissive and Declaratory, Expressive and Declaratory.

We would also have 10 combinations if we explored the possibility of combining three basic illocutionary acts:

> Assertive, Directive, and Commissive; Assertive, Directive, and Expressive; Assertive, Directive, and Declaratory; Assertive, Commissive, and Expressive; Assertive, Commissive, and Declaratory; Assertive, Expressive, and Declaratory; Directive, Commissive, and Expressive; Directive, Commissive, and Declaratory; Directive, Expressive, and Declaratory; Commissive, Expressive, and Declaratory.

If we accepted the possibility of a practice having four combined basic illocutionary purposes, we would have a logical space of five possibilities:

> Assertive, Directive, Commissive, and Expressive; Assertive, Directive, Commissive, and Declaratory; Assertive, Directive, Expressive, and Declaratory; Assertive, Commissive, Expressive, and Declaratory; Directive, Commissive, Expressive, and Declaratory.

We would also have the possibility of practices that mobilize all five distinct illocutionary purposes.

The fact that the communicational use of photographs can be related to the five basic illocutionary forces of John Searle's taxonomy allows us to consider that every photograph carries with it these five possibilities, which may manifest to greater or lesser degrees, depending on the context of use (in relation to a given practice) and/or the communicative intention of the user. It is important to recall that illocutionary purposes are general categories that encompass numerous acts sharing the same psychological mode and direction of fit, though varying in *force*—for instance, a request and a command are both directive acts, expressing a desire to bring the world into conformity with what is stated, yet they differ in the strength of that illocutionary force. Thus, we should *not* think of assertive, directive, commissive, expressive, and declaratory acts as "on/off" switches that can simply be activated or deactivated in the communicative use of photographs. Each of these categories involves continuous gradations of force that may vary across contexts and uses. Moreover, given that a single photograph can be used to perform different speech acts—potentially encompassing all five basic categories – the most appropriate metaphor for understanding photographic communication would be the following in Fig. 4.1.

The gradations of each *illocutionary force level dial* would determine the greater or lesser strength of the specific illocutionary force, potentially even being set to zero. And the combination of these five dials would determine the logical space of possibilities for the illocutionary purposes of different photographic social practices and for the various uses that photographs may assume within those practices. In simpler terms, assertiveness, directiveness, commissiveness, expressiveness, and declarativeness are forces that come in degrees and can be combined.

The idea that all five illocutionary force level dials can be engaged in a single use of a photograph is not far-fetched and can be illustrated by the example of an

Fig. 4.1 The five illocutionary force level dials of all photographs (image generated by AI)

advertising photograph. As we have seen on multiple occasions, advertising photographs aim to perform the directive act of prompting the viewer to carry out the action of purchasing the product (whether in a Gricean or non-Gricean way). But they also carry the burden of a commissive act, involving a negative deontological power: if the product is purchased, the user of the photograph is normatively bound to provide something that resembles what the photograph depicts. In addition, advertising photographs often perform expressive acts, manifesting (potentially through reflexive intention) a certain mental state in relation to the depicted object or scene. Furthermore, they may also perform a declaratory act, whose purpose is to establish the object as a symbol of social status; something, for instance, associated with wealth and success. And if the image is used also with the purpose of depicting a scene that actually took place, it may perform the assertive act of telling people how things are; carrying with it the commitment that the picture adequately fits the world. In this case, the photograph would exhibit gradations across all five dimensions of variation, with different illocutionary forces being attributed along each of the *level dials*. Each dimension would register a particular degree of force, depending on what the photograph depicts, how it depicts it, and how it is used within its communicative context and intention. The contextual elements related to far-side pictorial pragmatics would play an important role in determining the strength attributed to each of the possible illocutionary forces of photographic speech acts, as well as the commitments and requirements that follow from the assignment of specific forces to each speech act.

However, in scientific contexts it is possible that only the assertive dimension is mobilized. The more demanding the contextual requirements for precision and control over the elements of the picture and its mode of production, the more that particular *force level dial* is turned *all the way up*. In this case, the variation in force would correspond to the strength with which the image is used to assert that the world is a certain way—anchoring the strength of the force in the degree to which the practice of use ensures that the picture faithfully and accurately corresponds to reality. Consequently, the strength of the illocutionary force, once grounded in the fulfillment of the constitutive rules for assigning a status function to the photograph, would also determine the strength of the epistemic power the picture holds within the practice.

But there are artistic practices in which assertiveness may be set to zero, in the sense that there is no commitment whatsoever to the picture's correspondence with the world; it may even be explicitly recognized as a fictional image. In such cases, the entire practice may be structured around the expressive dimension: the greater the expressiveness, the higher the artistic and aesthetic value attributed to the picture within the practice. However, artistic practices can also combine different dimensions. One example is highbrow photojournalism, which brings together assertiveness and expressiveness, often aiming at the perlocutionary effect of sensitizing viewers to a humanitarian cause.

In other words, every photograph would carry within it the five *illocutionary force level dials* corresponding to the possible illocutionary forces, such that in a single image, one, several, or all of these forces might be attributed to varying degrees. A future finer-grained terminology would enable us to name and describe different gradations of variation within a single illocutionary dimension of photographic speech acts. And the combinatorial possibilities of the five illocutionary force level dials determine the logical space of photographic social practices and their communicative uses.

4.2 Felicity Conditions as Constitutive and Regulative Rules

The task of analyzing whether these combinations of the five illocutionary forces have counterparts in the real world is an empirical task that exceeds the scope of this book. What I want to emphasize is that, regardless of the number of illocutionary purposes a practice may have, when it possesses one or more illocutionary purposes, the structure of the practice can be understood in terms of the *conditions necessary for the successful and felicitous performance* of the act that constitutes the illocutionary purpose (or purposes). In other words, different practices may have as *constitutive* and *regulative* rules those elements that must be fulfilled for the illocutionary purpose of the practice to be achieved. Certainly, there may also be other types of rules that are independent of the purpose of the practice. However, the rules that will be important for constructing a social ontology of photographic practices will be those related to the felicity conditions of the illocutionary act that is the purpose of the practice.

As seen in Sect. 3.3, according to Searle (1969: 64–71), there are four types of *felicity conditions* related to speech acts:

- *Essential conditions*: these require that the illocutionary purpose of the act be intended by the speaker.

- *Preparatory conditions*: such as the speaker's authority and the existence of the extralinguistic institution.
- *Sincerity conditions*: these pertain to the possession of the necessary psychological states for the performance of some acts.
- *Content conditions*: these ensure that the *locutionary act* performed is appropriate so that, in that context, with the fulfillment of the other requirements, the illocutionary act is implemented.

The distinction between *constitutive* and *regulative rules* is proposed by John Searle (1969: 33–34). Regulative rules standardize practices that already exist, guiding how activities within these practices should be conducted. These rules have the imperative form "Do X." These rules govern behaviors in social interactions; such as the expectation to greet someone when introduced. While there is an expectation for participants to adhere to these rules, there is always a possibility of non-compliance; such as choosing not to greet someone upon introduction. Constitutive rules, in contrast, are foundational to defining what a practice is. These rules are so integral that deviating from them means not participating in the practice as it is defined. For example, in chess, moving a rook only forward, backward, or sideways is a constitutive rule. Moving it diagonally violates this constitutive rule, thus stepping outside the bounds of playing chess.

Although Searle's distinction between regulative and constitutive rules can be nuanced and doesn't always have clear boundaries (as noted by Raz 1975), parallels to this distinction can be found in the rules governing image use and production within photographic contexts. Various photographic practices have, over time, become sedimented and normatized, featuring rules that function as constitutive and/or regulative rules. This normatization in photography is teleological, oriented toward the illocutionary purpose envisaged by the practice and its perceived function.

In the realm of photography, constitutive rules are norms established within a photographic practice to support, enhance, or ensure the likelihood that photographs fulfill their intended purpose. The constitutive nature of these rules is such that non-compliance excludes the photograph from participation in that specific practice, thereby preventing the attribution of a status function and the powers that would follow from it (whether deontic, epistemic, or both). For example, in the context of documentary photography, since the purpose is to visually document and convey information about real events, epistemically reliable equipment must be used, and there is a prohibition against edits that could alter the pictorial content to the extent of changing the propositional content communicated. Thus, altering content-determining properties or staging a scene would breach this rule, disqualifying the image from being assigned the *status function* of a documentary photograph. Sanctions can be imposed (to the point of being legal sanctions) to ensure compliance with the rules.[16]

[16] Beyond the basic case of excluding the image from participation in the practice, a significant breach of constitutive rules could also give rise to a new practice or alter the understanding of what constitutes the practice.

Regulative rules, on the other hand, determine what is common or standard within a photographic practice, but breaking these rules does not preclude the image from participation. For example, in some practices, a regulative rule might be the use of a certain aesthetic or technique. Regulative rules can take the form "Do photographs with a certain type of content and/or aesthetic and/or method of production," "Use a certain type of equipment," and so on. Often, these rules are not explicitly formulated (verbally), but rather are the sedimentation of the habit that, in that context, certain equipment and procedures are used. And in the case of regulative rules, there is no requirement that non-compliance would exclude the photograph from participating in the practice. These rules regulate but do not constitute the practice. Deviating from these norms might challenge traditional techniques or established aesthetics but may not disqualify the work from being considered part of that photography practice. There are cases in which breaking regulative rules is even welcomed and can bring significant advancements to the practice.

However, the boundary between constitutive and regulative rules is not always clear in concrete cases, and there may be a degree of deviation from a regulative rule beyond which the photograph is no longer accepted within the activity. In such cases, the rule effectively functions as a constitutive one.

As previously seen, constitutive rules take the form:

- X counts as a photograph in practice P if...

What follows the conditional can be related to the four types of felicity conditions of speech acts proposed by Searle (1969, 1979):

- X counts as a photograph in practice P if X fulfills the essential condition, and/or preparatory conditions, and/or sincerity conditions, and/or content conditions.

The most restrictive practices will be those that have a greater number of conditions of the same type or more types of conditions as constitutive rules.

Rules can be either positive, outlining what actions should be taken, or negative, specifying what actions should be avoided or are forbidden. These rules can be explicitly stated or function implicitly as unwritten guidelines. Practices can range from being restrictive, with numerous constitutive rules, to more permissive, with fewer rules. For instance, assertive practices which are knowledge-based tend to be restrictive and involve many constitutive rules, whereas expressive practices, like artistic ones, are typically more permissive.

The determination of constitutive rules and their compliance is what allows the attribution of a *status function* to the photograph within conventional practices and its institutions.[17] The attribution of a status function would take the following form:

[17] As already mentioned, I also accept the possibility that a status function may be attributed by means of a declaration. (On this distinction, see Gouvea (2024)). However, I believe this is not the primary mode in ordinary *conventional* contexts.

- X counts as a photograph with a certain status function S in the practice P if X fulfills the essential condition, and/or preparatory conditions, and/or sincerity conditions, and/or content conditions.

As already mentioned, the attribution of a status function to a photograph frequently entails the granting of powers to the image within the practice (following the very model of deontic powers that Searle (2010) associates with the assignment of status functions). Photographs may thus acquire deontic powers—such as rights, duties, obligations, permissions, and authorizations—as well as epistemic powers.

The attribution of powers related to status function takes the form:

- If X has the status function S in the practice P, we accept that the photograph has power A in the practice P (the photograph does A).[18]

This paves the way for thinking about *factive experience* as a specific type of inference that can be applied to the use of the photograph, once it has been attributed a certain *status function* related to assertive practices, which, through constitutive rules, ensure that the photographic system operates as a reliable informational system. The factivity of photographic pictures can thus be understood as a positive *epistemic power* related to the assignment of a *status function*, through the fulfillment of conditions necessary for the performance of assertive acts.

We can modify the initial formulation in the following terms:

- If X has the status function S in the practice P, the fact that X means Y implies Y.

For example, by meeting certain requirements of a scientific practice, participants accept that the photograph holds the status function of a scientific photograph, which allows for factive inference. As a result, the photograph acquires the power to *prove* that something exists, to *attest* that an object has certain properties, or to *provide evidence* that a certain event occurred, among other things. These capacities of the image are epistemic powers tied to the attribution of a status function through compliance with constitutive rules, in a context that permits factive inference. But this kind of factive inference would not be appropriate for just any status function attribution. For example, in non-assertive expressive artistic practices, the attribution of a status function to a photograph as an *artwork* does not permit the inference "the fact that X means Y implies Y." As a result, the inference "the fact that X means Y implies Y" would not be applicable to these images.

Therefore, factivity does not belong to the *essence* of photography (as Robert Hopkins supposed in 2015, as seen in Chap. 1), but rather a specific type of inference about the world that becomes permissible once the photograph fulfills certain conditions related to the performance of assertive illocutionary acts, thereby

[18] Here, I am adapting to photography the structure of *the deontic power account* "We accept (S has power (S does A))," as presented in John Searle's theory, according to Åsa Andersson (2007).

enabling the assignment of a particular status function to the photograph.[19] And there are photographs that do not meet the conditions imposed by the practice, meaning that the power in the form "We accept that the photograph P has power A (P does A)" will not be attributed to the image.

But how can we think of the four types of conditions in photographic terms, as constitutive rules of photographic practices? Let us examine the conditions in more detail.

4.2.1 Essential Conditions

The idea of an essential condition has already been mentioned several times, as the fundamental condition for the performance of a speech act is that its illocutionary purpose be intended. In using the taxonomy of speech acts as a taxonomy of social practices, we can formulate the essential condition for participation in a practice as the constitutive requirement that the illocutionary purpose (or purposes) of the practice be intended. This condition is so fundamental that merely using a photograph within the practice brings about the assumption that the purpose has been intended for the use of the picture; thus, enabling the adoption of a non-intentionalist approach to communication within the practice, as the overtly manifest display of intention by the user of the photograph is not required.

The important aspect to emphasize is that intending an illocutionary purpose is articulated in relation to the *psychological mode* and the *direction of fit* of the specific illocutionary act (in acts where these elements are present). For an assertive act to be intended, there must be a *belief* that the content fits the world. For a *directive* act to be intended, there must be a *desire* for an action to be performed by the viewer/interlocutor, with the condition of satisfaction being the amendment of the world to what is intended. For a *commissive* act to be intended, there must be a *commitment* to a future action. For an *expressive* act to be intended, the *expressive* properties of the image must be recruited as part of a communicative intention. For a *declaratory* act to be performed, the picture must be used in such a way to bring about the alignment of the social world with what is depicted, having a double *direction of fit*. Consequently, in assertive, directive, commissive, expressive, and/or declaratory practices, participation in the practice requires that the use of photographs aligns with the psychological mode and the direction of fit (in acts where these elements are present) necessary for the intended illocutionary purpose.[20]

---

[19] These conditions will be described in Sect. 4.3 as content conditions concerning the production rules that a practice may impose as constitutive rules.

[20] Although in the case of declaratory speech acts no psychological mode is required, and expressive acts have no direction of fit.

Therefore, there can be no inconsistency between the engagement in the practice, intending its illocutionary purpose, and fulfilling the conditions related to the psychological state and the direction of fit (in cases where these determinations are required). For example, one cannot successfully and felicitously (in the technical sense) engage in a photojournalistic practice without the *intention* to perform the assertive act through the use of photography, which is part of the illocutionary purpose determined by the perceived function of photojournalism. Furthermore, it is not possible to intend to perform this assertive act without the *belief* that the depicted content *corresponds* to some extent with the world. Breaches in relation to the psychological mode can be understood as failures to meet sincerity conditions and can be addressed as part of an *ethics* of photographic communication. The use of a photograph in the context of photojournalism without the belief that its content corresponds to the world (for example, when the photographer knows that the image falsely represents the world or communicates a false proposition) would be at odds with the ethics of photojournalism, as it would contradict the purpose determined by the function of the practice.[21]

4.2.2 Preparatory Conditions

Many speech acts can be classified as *authoritative acts* because they demand, as a preparatory condition, the possession of a certain authority by the speaker and an appropriate relationship of this authority with the status of the interlocutor. There are several photographic social practices where the production and/or use of photographs within the practice require such preparatory conditions. In conventional contexts where photographs are assigned a status function that possesses powers requiring a high degree of epistemic rigor, it is quite common that only certified or formally recognized individuals can engage in these photographic practices. Medical practices based on photographic technologies, for example, where photography is granted the status function of a medical diagnostic imaging tool (and the image has the power to determine whether a certain treatment should or should not be administered), can only be performed by duly certified and formally recognized individuals. Without the required authority and the correct institutional linkage, the photograph will not be granted the status function and will not have the powers that would arise from this assignment.

It is important to understand how the status function and powers may depend on the fulfillment of the preparatory conditions. Let us consider the use of photographic pictures in scientific academic journals. These journals either are institutions themselves or are part of larger institutional bodies. The inclusion of photographic

[21] It is in this sense that I consider the theory proposed in this book to be a possible starting point for an ethics of photography.

images in such publications requires that the image be produced or used by someone with appropriate institutional credentials, such as affiliations with universities, research centers, or funding agencies. Typically, scientists submitting research for publication need to disclose the equipment and procedures used in producing the image.[22] Journals maintain specific protocols to ensure compliance with image production rules, including processes of blind peer review by experts in the field. Once these protocols are followed, the photographic picture used in the article is accepted for publication and receives institutional endorsement. Thus, it is the fulfillment of the conditions that enables the photograph to be accepted by the institution, thereby allowing, in that context, the attribution of the status function of a scientific photograph. With the attribution of this status function, it may be granted the epistemic power to *prove* the occurrence of a certain phenomenon, *document* the existence of a particular species, *support* or *challenge* a hypothesis, and so on. Disconnected from the institution and the authoritative figures associated with it, the photograph would not possess the status function or its associated powers.

But in practices with less rigorous epistemic standards, formal requirements of authority and institutional linkage are generally weakened. In some cases, authority is not based on formal certification but rather on institutional recognition that grants the individual the power to produce or use photographs with a certain status function. This would be the case in photojournalism, where formal certification (such as a university degree) is not frequently required, but rather the affiliation with an institution that recognizes the individual as a photojournalist.

Even in practices that are generally non-assertive, such as expressive artistic practices, formal certification is not required, but institutional recognition is. If someone is recognized by the institutions of the *artworld* as an *artist*, they become eligible to perform speech acts within those institutions, which can receive the status function of an *artwork*.[23]

In cases disconnected from extralinguistic formal institutions, the authority to perform certain acts is often bestowed by public opinion on someone based on the judgment of where one is located in relation to their various social identities (gender, race, social class, ethnicity, nationality, etc.). In these cases, the authority is related to the social recognition of the standpoint or positionality of the photograph's producer or user in society. This type of authority is generally connected to the performance of illocutionary acts in *non-conventional contexts* (which we will analyze in Sect. 4.4). For example, if the photograph is used to perform a speech act related to a specific gender/race, individuals recognized as belonging to that gender/race will have more authority to perform the act than those who are not. From this perspective, it is the possession of a certain biography that grants the authority required to perform felicitously the act.

---

[22] We will describe this kind of requirement in Sect. 4.2.4 as part of the *content conditions* determined by the practice.

[23] On this topic, see Abell (2012).

For this reason, especially in the case of art, a significant part of the artist's work extends beyond the artworks they produce and is related to the construction of a biography that authorizes them to perform the intended photographic speech acts. It is no coincidence that great artists often have significant and substantial biographies. In other words, the life one lives may grant one the authority to perform certain acts or prevent one from having the authority to perform certain acts.

The case of pornography, as interpreted by Rae Langton (which we have analyzed in several previous chapters), allows us to extend the idea of photographic authority to instances where no one explicitly performs the act or holds the authority to do so. Pornography performs the authoritative act of declaring women as socially inferior to men, even if there is no specific individual endowed with that authority. According to Langton, this authority derives from the role that pornographic pictures play in instructing their consumers on "which moves in the sexual game are legitimate" (2009: 45). In terms of Searle's taxonomy, we could say that the authority to perform the *declaratory* act of ranking women as socially inferior stems from the *directive* use of these pictures in instructing their consumers on how to act in sexual contexts. Authority, in this case, is a *consequence* of the directive use of these photographs and could, in turn, be sufficient to fulfill the requirement for performing the authoritative secondary act.

Similar cases can be found in many uses of photography where the figure of the photographer or image producer plays a peripheral or null role in the context of authoritative acts. A political propaganda campaign aimed at determining the social status of a politician may have the necessary authority to perform a declaratory act, even if we do not know who created or proposed the image, provided the photograph can persuade the public that this is the politician's social status. The authority required to fulfill preparatory conditions can be obtained through the persuasion of public opinion. We could also view these cases as an authority emanating from the collective that constitutes public opinion.

4.2.3 Sincerity Conditions

Many speech acts require certain psychological states in relation to the locutionary content of the act. If I perform an assertive act, I express *belief* in the content *p*. If I perform a directive act, I express the *desire* to align the world with the content *p*. If I perform a commissive act, I express my *commitment* to align the world with the content *p*. If I perform an expressive act, I *express* a mental state in relation to a certain content p. Possession of these psychological states constitutes a *sincerity condition* for the proper implementation of the act. This is reflected in the way the performative act would not be acceptable without the possession of the psychological state. For this reason, according to Searle, one cannot say, "I state that *p* but do not believe that *p*," "I promise that *p* but I do not intend that *p*," etc. (1979: 5).

The same applies to the use of photographs for performing speech acts that require the possession of a certain psychological mode or mental state. As seen in our analysis of essential conditions (Sect. 4.2.1), I could not use a photograph to *assert* something if I did not *believe* that the content fits the world. If I did not believe that the content of the picture fits the world, I could pretend to perform an assertive act, but it would be insincere. Similarly, I could not properly use a photograph to perform a *directive* act if I did not have the *desire* to align the world with the content of the picture. I could not effectively use a photograph to perform a commissive act if I did not have the *intention* to align the world with that content.[24]

In conventional contexts, since we can conceive of communication without the need for an ostensive manifestation of communicative intention, as intending the illocutionary purpose of the practice is a prerequisite for participation, the notion of sincerity conditions could be adapted to this type of situation. Just as the mere presence of a photograph in a practice allows us to attribute an intentionality aligned with the performance of the speech act that is the purpose of the practice, we can also, by the mere presence of a photograph in a certain practice, presume the possession of the psychological state necessary for the act's performance. And if it is later discovered that the producer or user was not in possession of the required psychological state, this could be understood as a violation of the sincerity conditions for engaging in the practice, which would render the speech act null.

But we can also consider sincerity conditions in relation to the performance of expressive acts. As I argued in Sect. 3.5, we can perform expressive illocutionary acts when the use of the photograph is intentionally related to its capacity to express certain mental states, and there is a *reflexive intention* that this intention be recognized by the spectator. This case differs from the perlocutionary use of expressive properties, where the mental state expressed by the picture or aroused in the viewer is not part of the user's reflexive communicative intention. In the use of photography as part of an expressive act, depending on the presuppositions that structure the context of use, the possession of the expressed psychological state could also be regarded as a sincerity condition for adequate performance of the act. For example, if I use a photograph of a beautiful sunny day at the beach with the communicative intention that the happiness expressed by the picture is felt by me and that this should be recognized by the interlocutor, then if I do not in fact possess this psychological state, the use is, in a certain sense, insincere.

4.2.4 Content Conditions

In the case of verbal speech acts, content conditions determine the requirements that the *locutionary act* must fulfill for the felicitous performance of the intended illocutionary act, provided the other conditions are met. In the case of propositional acts,

[24] Certainly, there is the possibility of communicative failures and infelicities. For example, even if I do not intend to give an order, my interlocutor may interpret my suggestion as an order. However, the fact that I did not intend it would contribute to the infelicity of the act (in the technical sense of the term used by John Austin (1962)). The typology of infelicities was addressed in Sect. 3.1.

the content conditions determine the type of proposition required for the performance of a particular type of act.

As we saw in Sect. 3.4, a photograph can play different roles in relation to speech acts. The photograph can serve as the proposition to which an illocutionary force is applied. Additionally, there are cases where the photograph, in its contextualized use, can itself determine or contribute to the determination of the illocutionary force. I will argue (in this section and in more detail in the following ones) that there are *content conditions* that may be imposed for a given photograph to serve as the proposition upon which a force is applied (whether verbally or presupposed in the context), as well as *content conditions* that apply to the photograph for it to determine or contribute to a specific illocutionary force.

We can consider content conditions in photographic terms as specifying the *type* of photographs that can be part of or determine a speech act. However, it is important to first note that photographs can be categorized into different *types* in at least two distinct ways. We can classify photographs into types based on their *pictorial content* or their *means of production*.

The type of a photograph can be determined in a pictorial sense, by what the photograph depicts and how it depicts it. A photograph that depicts a cat is of a different type than a photograph that depicts a dog. This is the descriptive sense that Nelson Goodman (1976) classifies as the determination of a *label* that designates the picture as an *image-of-a-cat* as opposed to an *image-of-a-dog*.

The determination of the type of picture could also involve its expressive properties in the following sense. A photograph of a landscape with nostalgic colors would be a different type of picture than a photograph of the same landscape with cheerful colors. Thus, an *image-of-a-sad-landscape* and an *image-of-a-happy-landscape* would be pictures of different types, even though what the picture depicts (the landscape) and its denotation (a certain place) could be the same.

However, the photograph's type can also be considered in relation to the *means of production*. Photographic pictures vary in types based on the photographic technology used and the material properties of the resulting image. In this sense, a digital photograph taken with a 300 mm lens and printed with mineral pigment on cotton paper is a different type of photograph than an analog 6x6 photograph taken with an 80 mm lens and produced through photochemical processes.

The determination of the type of picture in relation to the means of production plays a crucial role. Especially in assertive acts within epistemic practices, it is the regulation of the means of production that determines the degree of accuracy of the picture. As we will examine next, certain status functions are only attributed to pictures of a specific type that meet a certain degree of epistemic rigor in their production. But conditions on the means of production can also be found in non-assertive practices, where the image must be produced by specific methods. For example, one can only engage in artistic practices that recreate nineteenth-century production techniques if the photographs are produced in a very specific way.

Furthermore, as noted in relation to *photographic underdetermination*, visually identical photographs can differ in terms of their means of production. As a result, two pictures of the same *type* concerning descriptive content may belong to different *types* regarding their means of production. This, in turn, can affect the determination of the status function and the powers attributed to the pictures.

The thesis I here propose (which takes us beyond the mere parallel with Searle's notion of propositional content conditions) is that, for the performance of photographic illocutionary acts, it may be required that the photograph be of a certain type both in a *pictorial* sense and in terms of its *means of production*. These requirements can thus be imposed by a social photographic practice as part of the *constitutive rules* related to *content conditions* to ensure that the illocutionary purpose determined by the perceived function of the practice is achieved.

In our analysis of Robert Hopkins' use of Paul Grice's theory of communication (Sects. 1.2 and 1.3), we examined Sperber and Wilson's (1995) approach to communication. The critique of Grice allowed us to argue, based on Sperber and Wilson, that photographic communication could be composed of two layers: a foundational layer of information and a second layer indicating that the first layer was intentionally manifested. Regarding this structure of communication, we could conceive the idea of content conditions as the requirements that can be imposed on photographs for them to be used as the foundational layer of information in an ostensively intentional illocutionary act. In relation to social practices, we could conceive the idea of content conditions as the requirements that a social photographic practice may impose on the photograph for it to play the role of the foundational layer in an intentional act related to the illocutionary purpose of the practice. Thus, content conditions would dictate the type of photographic image that would be accepted as appropriate as locutionary act for achieving the illocutionary purpose and performing the speech acts relevant to the practice in question.

There are at least three types of rules that could be considered content conditions:

- Rules regarding pictorial content.
- Rules regarding means of production.

We could also include, as an extension of the idea of pictorial rules (which will relate to what is depicted and how it is depicted), the consideration of rules such as:

- Rules regarding aesthetics and style.

Since content conditions generally play the role of constitutive rules in conventional practices and are crucial for determining the status function of photographs and their deontic and epistemic powers, let us analyze in detail, in the next section, the elements that may be part of content conditions.

4.3 Pictorial, Production, and Aesthetics Rules as Content Conditions

As seen in Sect. 3.4, there are different ways in which photographs can be related to a speech act, playing different roles. We can think of photographic illocutionary acts with figurative pictures as having the form F(p), where "F" is the illocutionary force and "p" is the proposition upon which the force operates. The logical role played by the photograph in speech acts does not need to be restricted to the determination of a proposition "p" upon which the illocutionary force "F" is applied. It can also extend to include, in some cases, the establishment of the illocutionary force of the act itself.[25]

In the most basic cases, the photograph is used to determine a propositional content upon which an illocutionary force is applied. For example, one can point to a photograph and say "I promise to do this." The commissive act, in this case, is verbally added to the photograph as a means of determining a propositional content. The illocutionary force can be applied to the picture verbally or be presupposed in the context of use. For example, in a product catalogue, it is not necessary to verbally state that the company commits to providing an object similar to the one in the photograph if the viewer purchases the product, as this commissive act is an illocutionary purpose of the practice, presupposed by users of pictures in that context. Photographic communication in this case is structured around this illocutionary purpose, without the need for the illocutionary force to be verbally added to the photograph.

However, as seen in Sect. 3.4, there are uses of photographs where the role of the photograph is not merely propositional but also related to the illocutionary force itself. The way the photograph depicts an object in a specific context of use can perform a particular speech act in that context. It is the way products and food are depicted and the use of these pictures in advertising practices that would perform the directive act of prompting viewers to buy the product.

The proposed thesis is that, when a photograph plays a propositional role in speech acts within conventional practices, there are often *content conditions* connected to the *satisfaction conditions* of the relation between the *propositional content* and the *world*. These conditions determine the type of photograph that would be adequate for the type of speech act aimed at as the illocutionary purpose of the practice, and this will depend on the *direction of fit* that the picture has in relation to the world in the intended speech act. On the other hand, in cases where the photograph determines or contributes to the determination of the illocutionary force, there will also be *content conditions* that must be met for the illocutionary force to be

[25] As we saw in Kulvicki's analysis (Sect. 2.5), a photograph determines a proposition only when it is "embedded in a larger communicative act" (Kulvicki 2020: 48). The pictorial content of the image, in isolation from its use, is strictly descriptive. In order for a proposition to be determined, this descriptive content must be mapped onto a domain of existing objects. This, in turn, depends on various context-sensitive elements.

achieved through the picture and its use (although in this case they are rarely constitutive rules of the practice, but rather internal and necessary conditions for the picture to perform the intended illocutionary force). I will distinguish these cases as *propositional content conditions* and *illocutionary content conditions*, respectively. We will begin with the analysis of the former, and subsequently, we will examine the latter.

4.3.1 Propositional Content Conditions

Photographic practices with illocutionary purposes involving acts that have a direction of fit generally establish rules to ensure either the picture's correspondence with the world or the world's correspondence with the picture. Assertive practices have a *picture-to-world* direction of fit and often have rules aimed at ensuring (with varying degrees of rigor) that photographs are a truthful and accurate depiction of the depicted facts, to ensure the picture fits the world. In the opposite direction, directive and commissive acts have a *world-to-picture* direction of fit. Social practices that are primarily directive or commissive also often impose restrictions on the photograph to ensure that the world fits the picture. For example, advertising photography, although it may use mockups, models, and substitutes to depict the objects being sold, often has constitutive rules limiting the use of retouching or effects that may mislead consumers about the appearance of the product. These rules aim to ensure that the products purchased match the picture used as a model of the product.

The rules that govern the fit of the propositional content of the photograph in relation to the world (whether *picture-to-world* or *world-to-picture*) can be thought of as *propositional content conditions* imposed on photographs for their use within the practice. There are two basic types of rules that can be conceived in this sense of content conditions (although a third will be added shortly thereafter):

- Pictorial rules: which determine what the picture should depict and how it should be depicted.
- Production rules: which determine how the picture should be produced.

We can find a good example of *pictorial rules* within an assertive practice with a certain degree of epistemic rigor in the use of photographs for official documents. The photographs for identity cards in the State of Minas Gerais, Brazil, have the following explicit constitutive rules:

> The face and shoulders of the person photographed must be framed and centered. The eyes must be open, visible, and directed towards the camera. (...) The face of the person photographed must cover between 70 to 80% of the photo, from the chin to the forehead. Reflections and red eyes are not allowed. The photographed person must have a neutral expression or a discreet smile, in both cases keeping the lips closed without wrinkling the face. The use of glasses is allowed only when their non-use causes any awkwardness. In this

> case, the eyes must be fully visible, and the frame cannot be large, thick, flashy, or cover the eyes of the photographed person, even partially. In the case of a child, they must be without a pacifier, and the hands or any part of the body of the person holding them cannot appear in the photo. The use of heavy makeup or headgear is not allowed unless used for religious reasons, but even then, it cannot hinder the perfect visibility of the face of the photographed person. (Article 2 of Ordinance No. 2, of April 15, 2019, from the Identification Institute of Minas Gerais)[26]

These constitutive rules determine the content that the picture must have for the attribution of the status function of *photographs for identity cards* by the institutions related to the practice.

In addition to *pictorial rules*, practices with high epistemic rigor generally adopt as constitutive rules *production rules*. It is frequently the application of constitutive production rules that ensures the use of a photographic technology within the practice as a reliable informational system, allowing the attribution of a *status function* to the photograph that permits *factive inferences* of the type "If X has the status function S in the practice P, the fact that X means Y implies Y".

In forensic photography, for example, all stages of image production are explicitly regulated. As Edward Robinson (2016) states:

> The challenge for forensic digital evidentiary photography is that every digital image must be captured, stored, processed, and managed in a manner to ensure that it will be admissible in the courtroom. To meet this challenge, law enforcement agencies are required to have a Standard Operating Procedure (SOP) that describes the digital imaging process from the time the evidence is photographed at the scene until the processed digital image is presented in court. (2016: 631)

Deviation from Standard Operating Procedure would render the photograph inadmissible in the courtroom, thereby depriving it of the status function of evidentiary photography and the deontic and epistemic powers associated with that function. When a photograph is granted the status function of evidence in a trial, with its consequent powers (which also requires the fulfillment of other conditions, such as preparatory conditions of authority), the use of the image permits the use of factive inferences in that context.

We can think of these production rules as related to the four stages of photographic technologies described by Dominic Lopes (2016) in his version of the New Theory of the philosophy of photography. Thus, the rules could pertain to:

1. The pro-photographic scene
2. The light image
3. The photographic recording event
4. The marking of the surface

The elements of these stages that are generally regulated are as follows:

[26] Government of Minas Gerais. (2019). Retrieved from http://www.mg.gov.br/sites/default/files/servicos/arquivos/2019/portaria_2.pdf. Accessed on 12/01/2024. (Translation by the author).

1. Equipment: This pertains to rules about the type of equipment that should be used. These rules can specify which camera, filters, lenses, flash, and other photographic tools are appropriate or required for the practice.
2. Operation: These rules detail how the equipment should be operated. This includes aspects such as setting photometry, adjusting the depth of field, determining exposure time, and other operational parameters of the camera and associated equipment.
3. Recording: At this stage, rules dictate how the light-image of the photographic scene should be recorded. This might involve specifications about the type of film and film development processes to be used in analog photography, the type of sensor and digital format in digital photography, and the color system that should be employed. In digital cases, it may also include the type of compression accepted or the requirement for uncompressed files—such as RAW files—and metadata.
4. Marking of the Photographic Surface: These rules cover how the recorded information is to be processed and edited to produce the final picture. This includes the types of enlargement procedures, image development techniques, editing and digital printing procedures, and the materials to be used for printing and displaying the picture, such as different types of paper, chemicals, pigments, and electronic displays.

In restrictive epistemic contexts, with a high number of constitutive rules, such as medical diagnostic imaging, there are not only production rules but also rules to verify the proper functioning of equipment. It is not sufficient to use the right equipment correctly; it must also be assured that the equipment is functioning appropriately. This allows us to delineate the distinction between *first-level rules* about image production and *second-level rules* about the calibration of the accuracy and functionality of the equipment used.

Conversely, in permissive contexts with minimal production rules, only conditional factive inferences about depicted objects can be made. These inferences follow the logic of "assuming that such equipment and procedures have been used, the image might inform us that...". However, unlike more regulated contexts, there are no guarantees about the reliability of the information provided by the photograph. As a result, we would have no certainty about the epistemic powers that could be attributed to the picture. Without constitutive rules, there is a higher likelihood that the system is unreliable, so the fit between picture and world is not guaranteed, nor is there a clear determination of the type of information the photograph conveys or its accuracy.

The importance of determining the type of information that photographic technology provides can be noted in relation to the specification of the color system used for making inferences about the world. For instance, in the case of a photographic image containing only black, white, and gray, without knowing whether the system used is color or black-and-white, or if it is reliable, one cannot determine whether the grayscale has a syntactic function representing the colors of depicted objects. If the system is reliable and color-based (accurately mapping the colors of the depicted objects), a photograph depicting objects in black, white, and shades of gray implies that the objects themselves in the world are black, white, and gray. It would be necessary to know if the system is calibrated in order to make inferences about the exact tonalities of the objects depicted. However, if a black-and-white system is used, it is not possible to make inferences about the colors of the objects because the black and white system does not convey color information.

When a photographic system is familiar to us ("*entrenched*" in Nelson Goodman's sense (1976)), extracting information about the world becomes a straightforward and effortless process, providing us with a sense of transparency within the system. For instance, if one knows that a picture was produced with inverted colors within an RGB system and is acquainted with the system, it enables accurate inferences about the actual colors of the depicted objects, notwithstanding the color inversion present in the picture. A green apple in a photograph implies that the causal referent is a red apple. However, in certain specialized contexts, this familiarity often results from an extensive process of study and training, emphasizing the significance of understanding the rules of photographic production.

Once the practice has established pictorial and production rules, a characteristic aesthetic of the practice may emerge as a consequence of the use of specific equipment, operation rules, editing techniques, and so forth. Provided we are familiar with its characteristic aesthetics, this allows us to frequently use the aesthetics as a key to deduce which social photographic practice a certain picture belongs to. Thus, when we see a photograph that has the characteristic aesthetics of a certain practice, we can infer what illocutionary purpose was intended and the psychological mode and direction of fit associated with the practice. We can also assume the fulfillment of other additional conditions. But this presupposition can certainly be mistaken (as we will see in the next section). A picture can conform to the aesthetic standard of a certain practice without its constitutive rules being followed.

A characteristic aesthetic can also be adopted by a social photographic practice as part of the content conditions, functioning as a constitutive rule. Aesthetic rules are generally not explicitly verbalized and operate in relation to the aesthetic standards adopted by the practice. When deviation from the aesthetic standard prevents the photograph from participating in the practice, it can be said that the standard functions as a constitutive rule. When the deviation is undesirable but accepted, the rule is regulative.

- Aesthetic rules: which determine the aesthetic standard for photographs used in the practice.

The characteristic aesthetic of the practice is generally broad enough to allow for individual stylistic differences within its general aesthetic standard (although there are cases of use in scientific contexts where there is no room for individual stylistic variations). For instance, in humanistic photojournalism of the mid-twentieth century, black-and-white photography is commonly used to depict unscripted events and people in socially and humanitarian challenging situations. Viewers familiar with this genre of photojournalism, upon encountering a photograph with its typical general aesthetic, can quickly infer to which practice it belongs and the intended speech act. Aesthetics serves as a code, enabling the immediate identification of the type of speech act executed by the contextualized use of the picture. But although there is a typical general aesthetic, we can still distinguish the specific styles of Robert Capa, David Seymour, and George Rodger.

The style of execution of a verbal speech act is, according to John Searle (1979), a dimension in which speech acts may vary. Of the dimensions of variation proposed by Searle, the difference in style is one of the least developed and has a cryptic formulation in his 1979 book. The theoretical point he aims to make is that speech acts could have the same illocutionary purpose, direction of fit, force, institutional requirements, and speaker status, yet differ in their style of execution. The same type of act, therefore, could be performed in different styles, introducing a significant distinction in the actions performed. According to him, "'Announce', 'hint', and 'insinuate', for example, do not name types of illocutionary acts, but rather the style or manner in which a rather large range of types can be performed" (1979: ix). However, the text is not clear about how the difference between these acts would be a difference in the style of execution.

In photography, the idea that individual photographers possess distinct styles is well acknowledged. Two photographers holding equal institutional status might photograph the same scene with similar purposes, directions of fit, expressed psychological states, and general aesthetics. However, they could differ markedly in style, which in turn may impact the force of the speech act performed.

A photographer's established style can become a visual interpretive tool, allowing viewers to quickly grasp the communicative intention of a picture, and to situate it within a certain photographic social practice. Typically, a photographer creates pictures with consistent illocutionary purposes, directions of fit, psychological states, status, and institutional relations. Over their career, they might repeatedly execute similar speech acts through various photographs and projects. What changes is the propositional content, determined by the people and places portrayed. Recognizing a photographer's style can lead viewers to assume that other dimensions of variation align with those in the photographer's previous works, facilitating immediate comprehension of what is communicated (which speech act is performed).

4.3.2 Illocutionary Content Conditions

As mentioned earlier, the role played by photography is not limited to the determination of propositional content to which an illocutionary force is added. Photographs can, in some cases, determine or contribute to the determination of the illocutionary forces of the acts themselves. In such cases, we cannot think of content conditions as merely related to the satisfaction conditions for the fit of the picture with the world (whether *picture-to-world* or *world-to-picture*). We need to think of content conditions more generally, as related to the illocutionary force of the act. The content conditions, in this case, are the requirements that the content of the picture must meet for the force of the speech act to be achieved.

We can explore this idea by taking as an example the use of photographs in political propaganda, aimed at performing declaratory acts to determine the social status of the depicted person. A photographic campaign to portray a certain politician as confident, decisive, and competent, if it successfully convinces public opinion, can determine the social status of the portrayed person as confident, decisive, and competent. But to achieve this, certain content conditions must be met, although specifying these conditions is not a simple task. The picture must not appear excessively staged. The person portrayed must have a natural appearance, as if they truly possessed the properties attributed to them by the picture. The composition must be harmonious and balanced. The colors must also be balanced, and the lighting must be frequently soft and natural. The genius and skill of great photographers lie precisely in their ability to create pictorial contents that can determine a certain illocutionary force, thereby performing a particular speech act, even though determining the specific type of content that fulfills this function is not something for which there is an exact formula (unlike the determination of the fit between propositional content and the functioning of photographic technology as a reliable informational system).

Now contrast this with a political propaganda campaign that aims to negatively determine the social status of a rival candidate. In this negative use, the politician is generally depicted as aggressive, unbalanced, with an animalistic appearance. The lighting used is generally harsh, often with high contrast. The composition tends to be unbalanced, resembling pictures taken in chaotic situations. These different expressive properties (of the figure, the scene, and the design) are recruited as part of the pictorial content with the aim of performing the declaratory derogatory act.

We can then say that the photographs from the positive and negative campaigns have distinct contents, as they attribute different properties to the depicted subjects, and this specific content is a requirement for the execution of the intended act. Thus, it is a *content condition*. However, it does not function as a constitutive rule of the practice, but as a condition for the felicitous performance of the intended illocutionary act, since it has the power to determine the illocutionary force itself.

The same applies to many directive uses of photographs in advertising contexts, where the picture needs to have a certain type of content for the illocutionary force of prompting the viewer to want to buy the product to be achieved. An advertising photograph of food that depicts the food in an unappealing and disgusting manner would not fulfill the content conditions for the intended directive act.

In expressive acts, there are also conditions that the picture must fulfill, which are related to the illocutionary force. In the case of using photographs to perform expressive acts, the expressiveness employed by the user as part of their expressive act must be inherent in the content of the picture. Thus, the photograph must fulfill certain content conditions for the expressive act to be successfully performed. In almost trivial terms, this means that for a photograph to be used in an expressive act, the picture must itself be expressive, as it is the expressiveness of the picture that the user will depend on.

In sum: there are content conditions that concern the propositional role of the photograph in the correspondence between the picture and the world, and content conditions related to the role the photograph can play in determining the illocutionary force of the act.

We can summarize the ideas presented in Sect. 4.2 and its subsections as follows. There are different types of rules:

- Regulative and constitutive;
- Positive and negative;
- Explicit and implicit;
- First-level (about the four stages of image production) and second-level (for equipment calibration).

Contexts, in turn, can be distinguished between:

- Conventional and non-conventional;
- Restrictive and permissive.

The constitutive rules that serve as content conditions can be:

- Pictorial content conditions: about the correspondence of the propositional content of the picture to the world (in both directions);
- Illocutionary content conditions: about the content that the picture must have for the illocutionary force to be determined by the picture itself in the context of use.

Content conditions, especially in the case of pictorial content conditions, are rules that can be of the following types:

- Pictorial rules;
- Production rules;
- Aesthetic rules.

Furthermore, we can distinguish photographic systems regarding how photography provides visual information about the world in:

- Reliable informational systems and unreliable informational systems.

4.4 Photographic Communication in Non-conventional Contexts

Historically well-established social photographic practices have an illocutionary purpose determined by the perceived function of the practice, which fixes the primary speech act (or acts) performed by the photographs within that context. These practices frequently also have, in addition to regulative rules, constitutive rules to ensure that the illocutionary purpose of the practice is achieved. Practices that have constitutive rules can be described as *conventional*. Thus, based on the structure of social practices seen in the previous section, in conventional practices, photographic communication can be conceived in a way that mitigates the importance of Gricean reflexive intention, to the point of allowing the understanding of communication in a *non-intentional* manner. In conventional practices with constitutive rules, it is often unnecessary to overtly manifest which act is intended, as the mere use of the picture within the practice allows for the attribution of a communicative intention related to the illocutionary purpose of the practice. It is an essential condition for engagement in the practice that the illocutionary purpose(s) of the practice be intended. Thus, photographic communication in conventional practices would be non-intentional in the sense that the communicative intention aimed at realizing the primary speech act(s) of the practice could be equally attributed to all pictures used within the practice. This derivatively allows us to state that it is the photographic picture itself that performs the speech act in its contextualized use within the practice.

However, as mentioned before, there is a *continuum* of possibilities that vary between more restrictive and more permissive contexts, so the use of photographs can occur in contexts with a higher or lower degree of conventionality. At the opposite end of this spectrum, we can think of non-conventional contexts in which photography is used for communication without the prior specification of an illocutionary purpose or conditions to ensure that the purpose is achieved. How, then, is photographic communication structured in non-conventional contexts?

Let us consider an example to explore the challenge that this type of question poses. Consider a scenario where someone you have never seen stops you on the street and shows you a photograph. While you may recognize the pictorial content and formulate hypotheses about how the photograph was produced and what the communicative intention is, based on the picture's aesthetics and style, you would likely wonder, "Why is this person showing me this photograph? What does she want with it?" In technical terms, you are questioning, "To what illocutionary purpose is the use of this photograph intended?" In this case, understanding the illocutionary purpose will frequently necessitate a Gricean *reflexive intention* (analyzed in Sect. 1.3). Your interlocutor must manifest their communicative intention (often relying on verbal and additional information), you need to grasp this intention, and then notice how this intention is manifested through the use of the photograph. Thus, communication in this kind of context necessitates an exchange of hypotheses and assumptions about the communicative intention, which may need to be revised, refined, or abandoned during the conversation. As outlined by Sperber and Wilson, in this case, communication requires "an exercise in mind-reading, involving the inferential attribution of intentions" (2002: 3).

For photographic communication to be possible in a non-conventional setting, it is also necessary to determine the conceptual framework within which the proponent situates the picture, so that the context-sensitive determinations that fix the descriptive content and, depending on the use, the propositional content of the pictorial content can be provided. This pertains to the elements of *near-side pictorial pragmatics* that make it possible to use the picture as a basic layer of information (addressed in Chap. 2). Such determinations would already be fixed within a historically established conventional practice. However, in a non-conventional context, this will need to be established and recognized by the viewer. And photographic communication will only be possible if both the user and the viewer share, to a certain degree, the elements necessary for the determination of the descriptive or propositional content of the picture.

From the point of view of *far-side pictorial pragmatics*, in non-conventional contexts, the illocutionary purpose is not predetermined, and communication occurs without pre-established constitutive or regulative rules. Thus, it is not determined what the psychological mode and the direction of fit of the photograph's use are (in cases where such elements would be relevant). And, without the presupposition that some constitutive and/or regulative rules are in place, there is no prior determination of the type of information a photographic image conveys about the world, nor of its accuracy and reliability (since we do not know which content conditions the picture has fulfilled). Thus, we would not know what the epistemic powers of the picture are or whether the factive inference can be applied. Elements such as the psychological mode, the direction of fit, and the means of production would have to be made explicit by the user and recognized or understood by the viewer. This necessitates the viewer to make inferences and deductions, frequently in a hypothetical and conditional manner, which might be modified and reevaluated throughout the interaction. This use of photographs in non-conventional contexts will also be strongly based on the presupposition of Paul Grice's Cooperative Principle (1957), and communicative Maxims applied to photography—as we explored in Sect. 3.6.

Let us return to our example and imagine there are indications that the image shown by the person on the street is intended for an assertive purpose. If the image is assumed to be used for an assertive act, it is customary to presuppose that it has been produced in a manner that is epistemically reliable, complying with the Maxim of Quality—that the photograph should offer accurate visual information pertinent to the context (although we have yet to determine the degree of accuracy of this information). Consequently, the image is also expected to have a direction of fit from picture-to-world, making judgments about the truth or falsehood of its pictorial content pertinent as well (since assertive uses turn the context into a Quality Context, as discussed in Sect. 3.6). Other communicative maxims will also be presupposed—such as the Maxims of Quantity, Relation, and Manner. However, these assumptions might be re-evaluated during the interaction if it becomes apparent that the use is associated with an artistic practice not aimed at assertive/epistemic ends, leading to a reassessment of the presuppositions about its production, communicative intentions, and the Maxims in place. If the use is artistic without assertive intentions, this will imply a change in the psychological mode, the direction of fit, and the conditions the photograph and the user must meet for the speech act to be properly implemented (potentially even leading to the suspension of the direction of fit and the Maxim of Quality). If the communicative intention is instead directive or commissive, aimed at selling the depicted object, we would need to reassess these elements and reconsider the psychological mode, direction of fit, and conditions associated with this communicative use.

Frequently in such non-conventional settings, assumptions about the illocutionary purposes of a photograph, which dictate its production process and other conditions, are inferred from the picture's aesthetics. Various conventional practices with pre-established illocutionary purposes often exhibit distinct ways of depicting subjects and possess a characteristic aesthetic, allowing a competent viewer (someone well-versed in the history of photography and art) to infer the communicative intention and illocutionary purpose of the picture. The aesthetics of the picture also allow us to deduce which equipment was used and how it was employed, enabling us to infer whether the picture meets certain content conditions. Successfully interpreting these aspects requires a level of photographic literacy, where the viewer possesses a historically informed repertoire linking subjects, style, aesthetics, and communicative uses. Therefore, both the content of the image and its aesthetics can provide insights to a competent viewer about the picture user's communicative intention and its underlying illocutionary purpose. But these inferences may be flawed, and during the conversation, there may be a need to reassess initial hypotheses.

The idea of non-conventional photographic communication may appear far-fetched, but it is very close to the standard mode of communication within *contemporary art* contexts. When encountering a photograph on the wall of a contemporary art gallery, you will likely be able to recognize its pictorial content (if the photograph is not abstract). However, this marks the beginning of a *detective* process to deduce the artist's communicative intention. You may read the work's title and the curator's exhibition text. Understanding the artist's background (such as who they are and where they come from) may also be needed. You might also need to comprehend how the photograph was produced and its relation to the artist's body of work. In technical terms, these pieces of information constitute the *content conditions*, *preparatory conditions*, and *sincerity conditions* fulfilled by the artist and the photograph. These elements enable you to assess whether the intended speech act performed by the artist meets the requirements for felicitous implementation. If the intended speech act meets the requirements fulfilled by the artist and the photograph, the photograph may successfully perform the intended speech act, allowing us to attribute to the photograph the derivative power to perform that act. It is only through the determination of the communicative intention and the fulfillment of the necessary requirements for the proper implementation of the intended speech act that it will be possible to assign a certain status function to the picture and its respective deontic and epistemic powers.

We can illustrate the typical non-conventional photographic communication by taking as an example a photograph by Jeff Wall from 1992, "Dead Troops Talk (A vision after an ambush of a Red Army patrol, near Moqor, Afghanistan, winter 1986)", Fig. 4.2.

When examining Wall's work, we recognize an aesthetic typical of photojournalism, featuring elements of war photography. This aesthetic choice leads us to attribute to the photograph an assertive role, with the status functions of *documenting* atrocities and war crimes. The aesthetics of the scene and the figures suggest the desolation typical of war's aftermath. Noticing the meticulous composition, we might assume the image is the work of a skilled, accredited photojournalist, from a reputable international photo agency. On the basis of these assumptions, we are led

Fig. 4.2 Jeff Wall. *Dead Troops Talk* (A Vision after an Ambush of a Red Army Patrol, near Moqor, Afghanistan, Winter 1986), 1992. Transparency in lightbox, 229.0 × 417.0 cm. Courtesy of the artist (I express my sincere thanks to Jeff Wall for granting the rights to reproduce the picture. I am also grateful to the Gagosian Gallery, which represents the artist, for facilitating the licensing process and authorizing its use)

to treat the picture as factive, thereby applying the inference that "the fact that the photograph X means Y implies Y." However, upon closer inspection, certain elements seem incongruent with a war photograph; some figures display tragicomic expressions and behaviors at odds with the expected solemnity of such a context.

Investigating Jeff Wall's background, it becomes clear that he is not a photojournalist but a contemporary artist who often works with staged, large-format, digitally composed photographs. This insight demands a reevaluation of our initial assessment of the illocutionary purpose of the image and the communicative intention we attributed to the photographer. Reading of a curatorial text would reveal that this picture was crafted over 6 years in a studio setting, using actors and digital techniques, and is displayed as a large-scale lightbox artwork, characteristics aligning with artistic practices rather than photojournalism.

This revelation shifts our understanding of Wall's communicative intention. Rather than documenting war and its atrocities, his intention seems to be to challenge viewers and prompt them to reevaluate their immediate tendency to attribute an assertive illocutionary purpose based solely on the aesthetic and style of the photograph. Our initial assumption regarding the assertive and documentary nature of photography has proven to be mistaken. This recontextualization transforms the perceived speech act of the image from assertive to directive, wherein the communicative intention is to prompt the viewer to reassess their understanding of the communicative use of photographs—as if the photographer is saying, "Do not trust so eagerly the documentary nature of photography!" In more technical terms: "Do not trust your common-sense Orthodox intuition that photographs are factive!" As a result, with this *gestalt shift*, the image takes on new expressive properties, transitioning from tragic to tragicomic to ironic. The irony, in this case, also lies in the speech act performed by the artist through the use of a picture that leads us to mistakenly presuppose an assertive act, causing us to erroneously situate the picture within a journalistic/assertive social practice—resulting in a misattribution of its status function and epistemic powers.

The precarious nature of communication in non-conventional contexts, reliant on hypotheses and assumptions subject to revision, does not negate the necessity for rules and adherence to conditions, especially content conditions. Different photographic speech acts necessitate different conditions for their proper implementation, regardless of whether the context is conventional or non-conventional. For example, in a non-conventional context, as described earlier, where an unknown person shows you a photograph on the street, if the intended speech act is assertive, it is crucial that the image is produced in a manner consistently aligned with the intended purpose. The major difference is that in conventional contexts, the conditions for the felicitous implementation of the speech act are usually constitutive rules of the practice, which allow participants to assume that they have been previously followed by the photographs used within the practice. In non-conventional contexts, there is a need to somehow explicitly show that these conditions have been fulfilled.

In the realm of contemporary art, where experimentation with photographic production is common, it is not unusual for a photograph (or a series of photographs) to be accompanied by explanatory texts from curators or specialists. This aids in conveying the communicative intention of the artist, uncovering the illocutionary purpose being aimed for, and how it is manifested through the photograph. There is generally also a description of biographical elements and those related to the means of production of the image. Does the need for such explanations inherently indicate a failure in photographic communication? Should not the photograph communicate this on its own?

Contrary to the common-sense myth that good art requires no explanation, the necessity for additional contextual information often stems from the work's experimental nature, deviating from established practices and requiring a deeper understanding of photographic production techniques and art-historical knowledge for its full appreciation. Just as it is necessary to have a certain understanding of photographic practices to adequately grasp what the use of an image communicates within a given conventional practice, when encountering a new artistic work, we often need to acquire the knowledge required to comprehend the use of photography as proposed by the artist. To some extent, in contemporary art, each major artist inaugurates a new photographic practice or brings about significant transformations in already established practices. This demands that the viewer understand the purpose assigned to the photograph and the rules and conditions implemented and followed by the artist.[27]

[27] Furthermore, as seen in Chap. 2, it is necessary for the user and viewer to sufficiently share the conceptual framework that determines the pictorial content of the image and the mechanism of pictorial reference in use within the context. These determinations would pertain to the fixation of the context-sensitive elements of the *near-side pictorial pragmatics*.

Moreover, photographic communication adheres to the general communication maxims established by Paul Grice (1957). It is understood that photographic communication is fundamentally cooperative. In some contexts, it is expected that photographs provide true and relevant visual information, while in others, this may not be the case. Similarly, in certain contexts, we hold specific expectations regarding the causal genesis of photographs, but this may not be a universal requirement.

Throughout this book, we have embraced a *top-down* perspective on photographic communication, aiming to comprehend how social practices govern the speech acts performed through the use of photographs via the determination of constitutive and regulative rules. With a focus now on non-conventional contexts, we can examine photographic communication from a *bottom-up* perspective. We can regard photographic communication as starting from the determination of the preparatory, sincerity, and content conditions that the user or producer and the image fulfill. This determines which speech acts are available to be felicitously implemented through the use of the photograph. The user's or producer's background and biography are also crucial for the specification of context-sensitive determinations that fix the propositional/descriptive content of the pictorial content. Another key requirement involves understanding the psychological mode and/or direction of fit attributed to the photograph in its intentional use. This facilitates grasping the communicative intention that specifies the intended speech act. With this, we understand which illocutionary force is intended and which force is attributed to the photograph. Regarding the metaphor used in Fig. 4.1 (Sect. 4.1), based on the fulfilled conditions, psychological mode, and direction of fit, we can determine which *illocutionary force dials* the user aims to engage and the gradations attributed to them. The next step is to ascertain whether the image and the user meet the conditions for the appropriate implementation of the illocutionary force of the intended act (both from the perspective of propositional content conditions and illocutionary content conditions). This perspective allows for a different approach to understanding the dynamics of photographic communication.

Moreover, there is a *continuum* of possibilities between the extremes of conventional and non-conventional practices. Even in conventional contexts, there may be room for secondary speech acts that require an understanding of elements related to non-conventional communication, based on a Gricean reflexive intention, which may require a bottom-up approach.

References

Abell, C. 2012. Art: What it Is and Why it Matters. *Philosophy and Phenomenological Research* 85 (3): 671–691.

Andersson, A. 2007. *Power and social ontology*. Malmö: Bokbox Publications.

Austin, J. L. 1962. *How to do things with words*. London: Oxford University Press.

Goodman, N. 1976. *Languages of art: An approach to a theory of symbols*. 2nd ed. Indianapolis: Hackett.

Gouvea, R. 2024. Departing from Searle in the metaphysics of institutions. *Revista Dissertatio de Filosofia* 60:59–79.
Grice, P. 1957. Meaning. *The Philosophical Review* 66 (3): 377–388.
Hopkins, R. 2015. The real challenge to photography (as communicative representational art). *Journal of the American Philosophical Association* 1 (2): 329–348.
Langton, R. 2009. Speech acts and unspeakable acts. In *Sexual solipsism: Philosophical essays on pornography and objectification*, 25–64. Oxford: Oxford University Press.
Lopes, D. 2014. *Beyond art*. New York: Oxford University Press.
Lopes, D. 2016. *Four arts of photography*. Oxford: Wiley-Blackwell.
Raz, J. 1975. *Practical reason and norms*. Oxford: Oxford University Press.
Robinson, E. 2016. *Crime scene photography*. Amsterdam: Elsevier.
Searle, J. 1969. *Speech acts: An essay in the philosophy of language*. Cambridge: Cambridge University Press.
Searle, J. 1979. *Expression and meaning: Studies in the theory of speech acts*. Cambridge: Cambridge University Press.
Searle, J. 1995. *The construction of social reality*. New York: Free Press.
Searle, J. 2007. Social ontology and the philosophy of society. In *Creations of the mind*, ed. S. Margolis and S. Lawrence. Oxford: Oxford University Press.
Searle, J. 2010. *Making the social world: The structure of human civilization*. Oxford: Oxford University Press.
Sperber, D., and D. Wilson. 1995. *Relevance: Communication and cognition*. Oxford: Blackwell Publishers Ltd.
Sperber, D., and D. Wilson. 2002. Pragmatics, modularity and mind-reading. *Mind & Language* 17.
Tsohatzidis, Savas L., ed. 2007. *Intentional acts and institutional facts: Essays on John Searle's social ontology*. Dordrecht: Springer.

Afterword

The main ambition of this book has been to provide the basis for a theory of photography as a means of communication. This was achieved primarily by understanding how photographs are used to perform speech acts within social practices. This approach took us through an extensive theoretical exploration, proposing a distinction between two fields of study in the philosophy of photography. The first focuses on understanding the contextual elements that determine what a picture depicts and denotes, referred to as *near-side pictorial pragmatics*. The second, broader field, referred to as *far-side pictorial pragmatics*, focuses on the speech acts performed through the use of photographs and the principles and maxims of communication that govern their contextualized use. The primary contribution of this book lies in the development of a far-side pictorial pragmatics applied to photography, accomplished through the proposal of an ontology of photographic social practices.

But what practical insights can be drawn from the theoretical considerations developed here?

When confronted with a photograph, the starting point is to ask which social photographic practice the use of that picture is connected to. Placing a photograph within a social practice (whether through the aesthetics of the picture, its apparent method of production, as conveyed by the user, or through other means) serves a dual role.

Photographs are underdetermined from both a pictorial and denotative standpoint. Different contexts have different conceptual schemes, with varying levels of rigor, which determine what participants recognize as the pictorial content. Additionally, photographs can be used denotatively with varying levels of quantification, employing different mechanisms (such as iconography, generalization, and causal reference, among others). For a photograph to serve as a foundational layer of information in an interlocutory context, the user and the viewer must share sufficient agreement regarding the descriptive content and, depending on the use, the propositional content determined by the use of the picture in that context. It is also

G. Ghisoni da Silva, *The Philosophical Foundations of Photography as a Means of Communication*, Synthese Library 526,
https://doi.org/10.1007/978-3-032-02649-1

possible that there are context-sensitive determinations regarding the mental states, feelings, and emotions that the picture expresses, which can be recruited in communication.

However, there may be a profound difference between what the picture depicts and what its contextualized use intends to communicate. The same photograph can be used with different communicative intentions, and a photograph is not always used with the same communicative intention for which it was produced. The placement of a photograph within a certain social practice defines its purpose, based on a collective intentionality that determines the perceived function of the practice, as well as the requirements that must be met for participation in that practice. These elements would determine which speech act the use of the picture aims to perform. And based on the fulfillment of the requirements that constitute the rules of the practice, the photograph may be assigned a status function that endows it with epistemic or deontic powers.

In certain practices, only photographs produced or used by individuals duly authorized by the institutions that constitute the practice are granted a status function related to the practice. Moreover, only pictures produced in a certain way, depicting objects in a specific manner and adhering to a particular aesthetic standard, may be accepted. In technical terms, the conditions that a social practice may impose as necessary for participation, as part of its constitutive rules, can include essential, preparatory, sincerity, and content conditions. And the content conditions may include pictorial rules (concerning what must be depicted and how it is depicted), production rules (which pertain to the four stages of photographic technology), and aesthetic rules (concerning the aesthetic standard of the practice). The determination of these conditions is fundamental to photographic communication, as two visually identical pictures may have been produced by different means and fulfilled different preparatory conditions, thus possessing different status functions and powers. In other words, photographs are also *photographically underdetermined*. It is the regulation of production within the practice that determines what type of information about the world the image provides and with what level of reliability.

These contextual elements also influence the determination of whether, in its use, the photograph will be subject to the question of truth or falsity. Truth or falsity would be relevant to assertive acts, where the context of use could be described as a Quality Context. Thus, in non-assertive uses, a picture may not correspond to the world without being false (since it would not have a picture-to-world direction of fit). Different assertive practices will prioritize different aspects of the picture and will have varying levels of rigor in determining whether the picture corresponds to the world. Thus, a photograph that is accepted as true in one social practice may be considered false within another, which operates under different constitutive rules and a different level of epistemic rigor. But the most crucial point to emphasize is that photographs can be used to perform assertive acts with varying levels of quantification (whether the image is used to denote a type or iconographically), allowing us to differentiate between the truth of what the picture depicts and the truth of what it communicates. A truthful and genuine photograph from a pictorial and causal standpoint can be used to perform a false assertive illocutionary act (and vice versa).

In addition to assertive uses, the illocutionary purposes of practices may also be related to directive, commissive, expressive, and declaratory speech acts. In these different cases, the use of photographic images can have different directions of fit and require different psychological modes. Special attention was given to the fact that in directive and commissive practices, the image is used with a *world-to-picture* direction of fit (such that it is presupposed that the world must conform to the image, with a *future-oriented* intentionality). We also observed how, in expressive acts, we can distinguish the illocutionary act from the expressiveness of the image and its perlocutionary effects. Special attention was also given to declaratoryacts, in which photographs are used with the capacity, or in relation to the capacity, to determine the social status of the people or objects depicted. Photographic illocutionary declaratory acts are those in which photographic images are used to create or transform social reality, potentially leading to real-world changes as perlocutionary effects. This taxonomy of photographic speech acts facilitates the typology of photographic social practices through the idea that the purposes to which photography is directed within the practice constitute its primary illocutionary act.

These considerations lead us to conclude that photographs are also *communicatively underdetermined*, such that the same image could be used to perform different speech acts in different contexts of use. This would involve the attribution of different status functions to the image (depending on how it fulfills various conditions), resulting in different deontic and epistemic powers. And it is also important to note that the same photograph can be used to perform several distinct speech acts and be related to those acts in different ways, potentially even performing indirect acts. Thus, a single social photographic practice may encompass various speech acts as its illocutionary purposes, which form part of the essential condition for engagement in the practice.

Thus, we could say that every photograph carries with it the potential strengths of the five illocutionary forces, as represented by the *five illocutionary force dials* in the metaphor of Fig. 4.1 (Sect. 4.1). It is the context of use and/or the communicative intention of the user that determines which of the five will be activated and to what degree of strength.

When we include a photograph in a conventional social practice and all contextual requirements are shared by both the user and the viewer, we may have the *illusion* that the picture alone performs all the communicative work at both levels (of near-side and far-side pictorial pragmatics), leading us to attribute the performance of the speech act to the picture itself. This is evident in how, within a social photographic practice with constitutive rules, communication often does not rely on the reflexive intention. Intending the illocutionary purpose of the practice is an essential condition for participation and can be equally attributed to all images used within the practice (so that this communicative intention does not need to be explicitly manifested by the user). However, this carries the risk of leading one to believe that the photograph communicates on its own (as if the photograph could *naturally* communicate something). This is similar to the case of a native speaker who does not realize all that is necessary to know and share in order for the spoken sounds of that community to convey meaning and perform speech acts.

This communicative complexity becomes visible when we need to reconstruct photographic communication from its basic elements (*bottom-up*), such as when the picture is used without being tied to social practices with constitutive rules. This is a frequent case in the use of photography in contemporary art. Although the inclusion of photography within the context of contemporary art gives a general perceived function to which the purpose of photographs is linked, there is no prior determination of a specific illocutionary purpose nor the determination of constitutive or regulative conditions to follow. Contemporary art is a context of radical freedom, where part of the artistic value lies in how the artist constructs the necessary determinations for communication. The only general frequent requirement is the preparatory condition that the user or producer be recognized by the institutions of the artworld as someone authorized to use photographs with the assignment of the status function of *artworks*. But when faced with a photographic picture in a contemporary art setting, detective work is often required to piece together the understanding of which speech act the user aims to perform. We often need to know the title of the work, the artist's biography, how the picture was made, and so on. Sometimes, this understanding is only conveyed through the reading of a specialist's curatorial text. But does this mean that communication has failed? Shouldn't effective communication presuppose that nothing needs to be explained? Shouldn't the photograph be sufficient to communicate on its own?

The idea that successful photographic communication requires no additional information would amount to the following prejudice: only communication within practices I am familiar with, or by users I am acquainted with, constitutes genuine communication. In a trivial sense, if I do not speak the language or lack the conceptual requirements for a specific usage, nothing is communicated *to me*. But this does not mean that it does not communicate at all or that the failure lies with the speaker. It may be the case that *I* do not meet the requirements to understand what is communicated.

We can think of the use of photography in contemporary art in the following terms: each artist is proposing a specific photographic practice of their own creation or introducing variations on an established practice. It is due to the fact that many contemporary artists propose their own communicative photographic practice that, when those artists become famous and gain followers, a new photographic practice may emerge. It is therefore necessary to understand which conditions are met by the artist and the photograph (the preparatory and content conditions), as well as the communicative intention (the direction of fit and psychological mode), in order to determine the intended speech act. In sum, everything that is presupposed within an already established conventional practice must, in this case, be explicitly stated and shared between the user and the spectator. This additional information will be essential for addressing and overcoming the four forms of underdetermination: pictorial, denotative, photographic, and communicational underdetermination.

Finally, it is important to note one element that has permeated our analyses throughout the book. The perspective adopted here has prioritized the analysis of communication centered on the idea of the *use* of photographs. This has allowed us to understand how we can use the same photograph for distinct communicative purposes beyond the one for which it was produced. Photographic communication thus opens itself to the future, without the need for what is communicated to be anchored in the past or exhausted by the intentions and the original context of the image's producer at the time of its creation. What a photograph communicates can be renewed with each act of use.

Index

G. Ghisoni da Silva, *The Philosophical Foundations of Photography as a Means of Communication*, Synthese Library 526,
https://doi.org/10.1007/978-3-032-02649-1

MIX
Papier aus verantwortungsvollen Quellen
Paper from responsible sources
FSC® C105338

If you have any concerns about our products,
you can contact us on
ProductSafety@springernature.com

In case Publisher is established outside the EU,
the EU authorized representative is:
Springer Nature Customer Service Center GmbH
Europaplatz 3, 69115 Heidelberg, Germany

Printed by Libri Plureos GmbH
in Hamburg, Germany